Françosi Sigismond Jaccoud

The Curability and Treatment of Pulmonary Phthisis

Françosi Sigismond Jaccoud
The Curability and Treatment of Pulmonary Phthisis
ISBN/EAN: 9783337140755

Printed in Europe, USA, Canada, Australia, Japan

Cover: Foto ©berggeist007 / pixelio.de

More available books at **www.hansebooks.com**

THE
CURABILITY AND TREATMENT
OF
PULMONARY PHTHISIS.

By S. JACCOUD,

PROFESSOR OF MEDICAL PATHOLOGY TO THE FACULTY OF PARIS; MEMBER
OF THE ACADEMY OF MEDICINE; PHYSICIAN TO THE LARIBOISIÈRE
HOSPITAL, PARIS; ETC.

TRANSLATED AND EDITED BY

MONTAGU LUBBOCK, M.D. (LOND. AND PARIS), M.R.C.P. (ENG.),

ASSISTANT PHYSICIAN TO CHARING CROSS HOSPITAL, AND TO THE HOSPITAL FOR SICK
CHILDREN, GREAT ORMOND STREET, LONDON.

NEW YORK:
D. APPLETON AND COMPANY,
1, 3, AND 5 BOND STREET.
1885.

PREFACE.

M. JACCOUD, the eminent Professor of the École de Médecine, Paris, is generally recognized on the continent as one of the best authorities on pulmonary phthisis, and it has been thought that an English edition of his work would be acceptable to those interested in the subject. His views are stated with so much clearness and the reasons for their existence so fully set forth in the work itself, that it is scarcely necessary to say anything by way of introduction. I may, however, perhaps add a few words; firstly, with reference to the pathology of the disease; secondly, from the fact that the bacillus, whose presence is regarded by many as the characteristic, and even the cause, of tuberculosis, was only discovered by Dr. Robert Koch in 1882, namely, at a date subsequent to the publication of this book.

With respect to the pathology of pulmonary phthisis, the author's ideas, though analogous to those which have been and are still held by many of the best authorities, are expressed in terms which necessitate a knowledge of the different opinions expressed in recent years upon this subject. It will be remembered that it was not until the eighteenth century that phthisis was specially ascribed to pulmonary lesions, and was regarded as presenting several distinct forms which were named, partly from the supposed cause, partly from the symptoms of the disease. In 1820 Laennec maintained that the different kinds of pulmonary phthisis were dependent pathologically upon a different degree of development in the lesion, this being the doctrine of "unity" mentioned by the author of this work. At the same time it was believed that "tubercle," as the lesion was named, might assume two special forms:—

1. That of isolated nodules.
2. That of an infiltrating mass.

In 1850 Reinhardt affirmed that the infiltrating masses of so-called tubercle consisted in reality of pulmonary alveoli filled with epithelial cells, and blood corpuscles as in catarrhal pneumonia; that these became granular and underwent fatty degeneration as in the last stage of that disease; and that tubercle was in this case but the result of ordinary pneumonia, in which the lung could not divest itself of the inflammatory products. Hence the term "caseous pneumonia," as applied to this form of tuberculosis; namely, to the disease which produces the massive tubercle, the tubercular infiltration of Laennec, the pneumonic tubercle of Grancher, the tubercular agglomerate nodule of Charcot as contrasted with the tubercular granulation of Virchow, the grey granulation of Bayle and Laennec, the "tubercule granulique" of Grancher, the miliary tubercle of Laennec, the tubercular peribronchial nodule of Charcot.

Virchow, believing the elements of the grey granulation to be derived from a special proliferation of the cells of connective tissue, denied the relation supposed to exist between miliary tubercles and caseous infiltration; the latter, in his opinion, being due to catarrhal pneumonia. Hence the cells which accumulate within the pulmonary alveoli were regarded as mainly, if not entirely, due to epithelial proliferation. Thus, in the opinion of Reinhardt and Virchow, the caseous infiltration of so-called yellow, cheesy tubercle, differs anatomically from the grey granulation; the former being due to inflammation, and produced by caseous pneumonia, while the latter is a living growth of which the structure is almost identical with that of adenoid tissue, and therefore to be looked upon as a lymphatic tumour. Such is the dualist doctrine as distinguished from that of unity promulgated by Laennec.

Grancher* is believed by M. Jaccoud to have proved, in 1872, that, from an anatomical point of view, tubercular granulations on the one hand, and the so-called pneumonic infiltration on the other, have the same structure; and this idea is represented as being confirmed by other distinguished observers.

Clinically, the author admits two distinct varieties of the disease: one the inflammatory, or pneumonic form; the other the chronic, or ordinary form of the complaint. Thus clinical duality is supposed to coexist with pathological unity.

Secondly, with respect to the micro-organism found in tubercular lesions. "The bacillus tuberculosis," says Dr. Green, in his work upon Pathology and Morbid Anatomy, "is the cause of all tubercular processes;" and though it is by no means universally acknowledged † that this is the actual

---

* "De l'Unité de la Phthisie." Thèse de Paris, Fevrier, 1873.

† Thus Dr. Henry MacCormac observes: "Consumption, human beings regarded, is not communicated by inhalation, nor yet, I submit, by infection. Dr. Koch's bacilli do not occasion phthisis; it is phthisis which occasions them;"

cause of the disease, there are strong reasons for such a conclusion. The idea that tuberculosis is an infectious disease has been long held. The author of this work was of that opinion, and believed the infective agent not only to pass in some way from person to person, but that auto- or self-infection was possible, the virus being disseminated through the body from pre-existing caseous or tuberculous deposits which contained it. The nature of the agent was, however, quite unknown to him, nor could he describe its size, conditions of life, etc., as can now be done. Nor was the presence of such a contagium considered necessary. Malnutrition, understood in its most general sense, was the chief factor in the liability to tuberculosis, and the actual production of tubercle, whether it be an exudation or cellular formation, was regarded as due to an irritative process termed "phymatogenous (tubercle-producing) irritation," or to actual inflammation of the lung. Though regarding every form of phthisis as curable, he looked upon the so-called inflammatory forms as specially so by means of fibrous transformation of the tubercle, the result to be specially sought in all cases of this disease.

With respect to infection, the disease was supposed to be communicable by means of inoculation, the inhalation of air containing particles derived from a tuberculous patient—the milk and possibly the meat of cows affected by tuberculosis—

and on the title-page of his remarks upon "The Ætiology of Phthisis" is the following chorus of the possible supporters of Dr. Koch :—
"What is consumption? The bacillus.
What is the bacillus? Consumption.
But what causes consumption? Why, the bacillus.
But what causes the bacillus? Consumption."

so that the treatment should be regulated accordingly. At the same time, tuberculosis was looked upon as the least infectious of all transmissible diseases, special susceptibility of the exposed organism being required, which he named by the term "receptivity" (*receptivité*). The character, however, of the infective agent was unknown, nor was it regarded as a cause of the "phymatogenous irritation" to which tuberculosis was supposed to be due.

M. Jaccoud's reputation is justly so great that his opinions with respect to the treatment, whether hygienic, medicinal, by mineral waters, or climate, will be read with general interest.

The names of the principal books which contain accounts of the places visited by consumptive patients, and to which allusion is made in this book, are mentioned in corresponding notes.

MONTAGU LUBBOCK.

19, GROSVENOR STREET, LONDON,
January 1, 1885.

AUTHOR'S PREFACE.

THIS book contains the substance of lectures delivered in December 1880 and January 1881; in the first chapter will be found an account of the reasons which led to their delivery, reasons which also explain why this work has been published.

Such a decision was justified both by the original character of certain pathological views, and by the novelty of my conclusions and methods of treatment. The latter were new, whether the prophylactic plan be considered or the four methods ordinarily adopted, namely, the hygienic method—in which I include hydropathy (*hydrothérapie*) and aërotherapeutics (*aerothérapie*)—the treatment by means of drugs, that by mineral waters, and the climatic method. With regard to the thermal and climatic modes of treatment, my conclusions are not only new, but have also the advantage of being based upon personal acquaintance with the places mentioned.

Such were the reasons for undertaking this work, which seemed to me not only valid, but even obligatory.

S. JACCOUD.

CONTENTS.

CHAPTER		PAGE
I.	On the Curability of Phthisis	1
II.	Conditions which influence the Curability of Pulmonary Phthisis	29
III.	Conditions which influence the Curability of Pulmonary Phthisis (*Continued*)	46
IV.	Prophylactic Treatment	61
V.	Prophylactic Treatment (*Termination*)	87
VI.	Treatment of the Ordinary Form of Phthisis	119
VII.	Treatment of the Ordinary Form of Phthisis (*Continued*)	149
VIII.	Treatment of the Ordinary Form of Phthisis (*Continued*)	184
IX.	Treatment of the Ordinary Form of Phthisis (*Termination*)	214
X.	Treatment by Mineral Waters	239
XI.	Climatic Treatment	283
XII.	Climatic Treatment (*Continued*)	313
XIII.	Climatic Treatment (*Conclusion*)	354
	Index	399

THE CURABILITY AND TREATMENT

OF

PULMONARY PHTHISIS.

CHAPTER I.

ON THE CURABILITY OF PHTHISIS.

Introduction—Reasons for the work—Pathogenic principles (*Principes de pathogénie*)—Therapeutic consequences.
Anatomical unity of phthisis—Writings of Grancher, Thaon, and Charcot—Clinical duality of the disease—Connection between doctrine of unity and notion of curability.
Curability of tubercle; curative process—Recovery from phthisis—Relative and absolute recovery—Anatomical and clinical proof—Error in the prognosis as regards cavities in the lungs—Variable signification of this lesion.

THE treatment of pulmonary phthisis will form the subject of this work. Convinced and certain as I am of the curability and possible arrest of this formidable disease, I believe it my duty to communicate in every detail the results both of my studies and experience. It is therefore my intention to consider at some length the question of its treatment, doing this for reasons of different kinds which I would first explain.

In the first place, the undoubted importance of this subject causes it to deserve constant consideration on the part of the physician, the value of this reason being truly measured by the frequence and severity of the disease. This fact is so evident that to consider it at greater length would be quite unnecessary.

The second reason, having a personal character, though equally valid, is altogether different, as will be understood.

The most important question in the treatment of phthisis is recognized to be the interesting and complex problem of climatic stations in winter or summer. There is no other disease in which the climate may be regarded to the same extent as a truly therapeutic means of treatment. It is equally powerful to do good or harm according as it is rationally or irrationally applied, and properly or improperly adapted to the indications furnished by the patient. With regard to this subject I believe myself to have acquired special knowledge.

Studies continuously pursued during many years have enabled me to know, by personal investigation, all places which are of special benefit in the treatment of phthisis. I have seen, and in many cases frequently, all such places in Europe, including those of Greece,* Sicily,* and Norway;* I have studied with care the celebrated station of Davos* in the midst of winter, since a visit at any other season would be totally useless; I have seen the stations in Asia; and, lastly, have visited Madeira,* the Canary Islands,* and the principal stations of Morocco* and Algiers.* Thus I am not unreasonable in claiming special acquaintance with this subject.

An immediate acquaintance with the places inhabited is the indispensable condition of really knowing them, whether with the view of teaching others, or for medical practice, since even the most complete knowledge of what has been written cannot supply the valuable information derived from personal investigation. Doubtless by sufficient study a more or less complete notion of the climatic conditions of the different places may be acquired, and I would even say that the knowledge

* See chap. xi. and xii.

thus obtained is indispensable in choosing a suitable residence; it should not, however, be supposed that the meteorological data supplied by books, which can be read at a distance, supply all the knowledge requisite in order that a physician may make his choice. Such would be a most mistaken idea, since many other things of equal importance must be taken into consideration. It is quite as indispensable, if not more so, in the interest of the patient, that the physician should have complete knowledge of the sanitary and hygienic conditions of any place which claims to be a "medical station;" and that he should know the true resources of the station with regard to residence, food, hygienic and therapeutic requirements. Whence can the knowledge of these facts be obtained, so important as to surpass all else in my eyes, as it should in the eyes of all, so well as from personal observation?

More than this, and with respect to the climate properly so called, information derived from writings cannot take the place of ideas furnished by direct investigation, for the following reasons. Two places possessing the same climate may have a very different medical value, and this solely on account of their topographical disposition. What would the meteorological tables say on this subject? Nothing. Again, the impression made upon the organism by different climates, is to some extent independent of the certain and describable phenomena which constitute the climatology of the region investigated. With similar mineral waters, barometrical, hygrometric, and anemological (*anémologiques* *) conditions, two places may still be most dissimilar from each other in their effects either upon the healthy person or the invalid. It is vain to seek in the table of atmospheric changes for the explanation of that special impression which the climate

* The word "anémologiques" (ἄνεμος, wind; λόγος, account), meaning connected with the wind, has been translated literally, since the word "anemology" exists in the English language. See New Sydenham Society's "Lexicon of Medicine."

makes upon the affected organism—an impression which, in fact, is indefinable, and of which the cause cannot be ascertained. It would only be discovered in the place itself, and must be felt to be understood,—a fact which constitutes one of the fundamental elements of the problem.

I do not insist upon this point, its truth being already known, and the statement one which cannot be denied. In order to decide between climatic residences, in order to give or deny to them the character of medical stations, and to make a rational choice between them, appropriate to the general and special conditions of the patient, the places must themselves be visited. Though some other course may be usually adopted it is certain that nothing in this matter can replace personal investigation. The knowledge thus obtained gives to the medical opinion a precise and substantial character which is of peculiar advantage in this disease. Having obtained, and not without trouble, such increased knowledge, I would impart to others, as far at least as the subject permits, the advantage derived from it. Such then is my second reason.

I have still a third one. For more than fifteen years I have unceasingly studied the questions connected with the climatic treatment of phthisis, and it would, I think, be right to make known to others the conclusions at which I have arrived. Of these I am the more ready to speak, since they differ materially from the somewhat routine rules which guide even now the practice of most physicians.

The truth is, that in consequence of these studies I have been led to make considerable and precise restrictions as to the use of the more southerly stations, to fix upon new principles the *general* indications for warm climates, and to establish upon a fresh basis the *particular* indications of the different climatic groups. It appeared to me, therefore, that it would not be fruitless to consider these questions whose solution is too often marred by the preoccupation of local interests.

No less useful would it be to speak of mineral waters, as used in the treatment of pulmonary phthisis, and to give my views and personal experience on that subject, feeling that I can speak with the same competence of this plan of treatment.

I would also seize this opportunity for the purpose of fixing the attention upon the importance of *hygienic treatment*, and of certain therapeutic methods which seem to be too much neglected, and specially in this country. It is my intention to speak of the *hydrotherapeutic*,* *aërotherapeutic*,† or *pneumatic* treatment, and of the cure by milk. I shall also mention the result of my studies with regard to a method highly spoken of in recent times, namely, the treatment by *inhalation of benzoate of soda*, and of the effects of *salicylic acid* used as a febrifuge.

I would also use this occasion for another purpose, namely, that of affirming once more, with a conviction strengthened by long experience, certain principles which I have laid down since my first work was written on this subject, and which I have some right to claim as my own. These principles refer to the general pathology of tubercle and phthisis; but contain special allusions to the therapeutical aspect of the disease, and, consequently, have a legitimate place in our present studies, which it would be best to commence with their consideration.

This subject was first considered by me in 1862. I stated at that time that tubercle always results from imperfect

---

* Hydrotherapeutics (*hydrothérapie*—ὕδωρ, water; θεραπεία, treatment) is the treatment of the disease by means of baths, etc., as considered in chap. v. and vi., etc.

† Aërotherapeutics (*aérothérapie*—ἀήρ, air; θεραπεία, treatment) is the treatment by means of air, etc., as considered in chap. v. and vi. The word *pneumatic* (πνεῦμα, air) is used with the same meaning.

nutrition, differing thus both from Graves,* who saw in it a product of nutrition pathologically perverted by scrofula, and from Hughes Bennett,† who believed some disturbance of the digestive functions to be the cause of tubercular formations. These two distinguished physicians, from their point of view being too exclusive, saw but one possible origin of the disease, and thus placed its ætiology within limits which were too narrow to be generally respected. My proposition with its general formula included all the conditions which, through different organs and by different processes, might give rise to defective nutrition, or, as I named it subsequently, *constitutional hypotrophy and dystrophy* ‡ (*l'hypotrophie et la dystrophie constitutionelle*). Numerous indeed are these conditions, and in such a vast sphere I was equally far removed

---

* It was in 1862 that Professor Jaccoud's translation of "Clinical Lectures on the Practice of Medicine," by R. J. Graves, was published, with notes written by himself. Graves believed the debilitated state of constitution which is termed the scrofulous habit to be the origin of tuberculosis (*Ibid.*, vol. ii. p. 90), and thus he states (p. 92) : "We have three distinct forms of disease in the lungs all arising from scrofula, namely, scrofulous pneumonia, scrofulous bronchitis, and tubercular development."

† Hughes Bennett ("The Pathology and Treatment of Pulmonary Tuberculosis," Edinburgh, 1853) believed disease of the "primary digestion" to be the origin of phthisis. Supposing that tubercle was an exudation of the liquor sanguinis (chap. i. sect. 2), he says (chap. i. p. 13), "Now, as the nutritive properties of the blood are entirely dependent on a proper assimilation of food, and as this assimilation must be interfered with in the morbid conditions of the alimentary canal, the continuance of such conditions necessarily induces an impoverished state of that fluid, and imperfect growth of the tissues. When under such circumstances exudations of liquor sanguinis occur, they are very liable to assume the form of tubercles, and if they are poured into the lungs, there are then produced those changes and that condition which have been denominated by the German pathologists pulmonary tuberculosis."

Hughes Bennett makes a similar assertion in his "Clinical Lectures on the Principles and Practice of Medicines," pp. 741, 742, concluding it with the remark, "From a study of the symptoms, causes, morbid anatomy, and histology of *phthisis pulmonalis* we are therefore led to the conclusion that it is a disease of the primary digestion."

‡ The words "hypotrophie" (ὑπό, under, indicating a deficiency, and τροφή, nourishment), and "dystrophie" (δύς, with difficulty or ill, and τροφή, nourishment) would mean "deficient and improper nutrition."

both from the too exclusive conception of Graves as respects scrofula, and from the equally narrow idea of Bennett with regard to the process of digestion. There is not, in fact, one of the multiple and diverse acts of the function of nutrition which might not, if persistently disturbed, produce that nutritive defect which characterizes the tubercular diathesis. Digestion, absorption, assimilation, hæmatosis, are all elements converging towards the same result, which is nutrition, and each of these can interfere with equal power in bringing about that hypotrophy which may produce tubercles. To leave no doubt as to the general meaning which I gave to the expression "imperfect nutrition," I also laid down the following proposition in the first edition of my treatise on Pathology in 1869:—*The tubercular diathesis is essentially constituted by insufficient nutrition* (insuffisance de la nutrition), *this term being understood in its widest physiological sense.*

Though since the year 1862 I have looked upon constitutional debility resulting from imperfect nutrition as the chief cause of tuberculosis, I showed at that time that, as regards its mode of production, tubercle bears a decided resemblance to inflammatory products. Thus, whatever histogenetic theory (*théorie histogénique**) be adopted with regard to its production, exudation or cellular formation would be but the final result of an irritative process which I have termed *phymatogenous irritation* (*irritation phymatogène* †). I remarked in the same work, in 1862, that these new views as to the origin of tubercle

---

* The term "théorie histogénique" (histogenetic theory—ἱστὸs, tissue; γένεσις, origin) implies the theory which regards the origin of (tubercular) tissue.

† *Irritation phymatogène.* These words mean "tubercle-producing irritation." (φῦμα, tumour or tubercle; γεννάω, I produce). Gueterbock gave the name of "phymatine" to an organic substance which in his opinion was peculiar to tubercle, and Lebert applied the term "phymatoid" (φῦμα, tubercle; εἶδος, appearance) to that condition of morbid growths in which the tissues were of a yellowish-white colour analogous to that of tubercle.

condemn both the theory of epigenesis * (*épigénèse*) maintained by Laennec, and that of heteromorphism † (*hétéromorphisme*), while they explain a fact of great practical importance, namely, the influence which inflammation has upon the development of tubercles.

I will recall the words of my statement so as to give to this retrospective view all the precision requisite. "Inflammation affecting the respiratory organs," I said in one of my notes to the Clinical Lectures of Graves,‡ "has in this case a double mode of action; at one time it hastens the evolution of pre-existing tubercles, at another it favours the development of these abnormal products where they had remained until then innocuous in the lungs. The first effect never has been and never can be denied; but the second has given rise to numerous discussions, which are not even now settled. This necessarily occurred, for so long as tubercle was regarded, as it has been by Laennec, as a new tissue, really living and possessing in itself the cause of the changes which it undergoes, it was impossible, even though individual predisposition be supposed of great importance, to perceive the slightest relation between the existence of pneumonia for instance, and the appearance of tuberculous tissue of special structure. This difficulty exists no longer, and we can now easily understand how pneumonic exudation in a predisposed subject may be modified owing to this predisposition in its most intimate composition, and may also deviate from its natural evolution. We can delay, and pos-

---

* According to Laennec, tubercle is not a simple transformation of degenerate matter, but a new and accidental tissue without analogous structure in the healthy body, and developed in its entirety by epigenesis (*épigénèse*, from ἐπὶ, upon; γένεσις, development) in the midst of displaced, but not destroyed organs.

† The doctrine of heteromorphism (ἕτερος, different; μόρφη, form) also supposes tubercle to be composed of elements different from those which normally exist in the body.

‡ R. J. Graves, "Leçons de Clinique Médicale." Ouvrage traduit et annoté par le Docteur Jaccoud, 1862, tom. ii. pp. 142, 143.

sibly avert, the formation of tubercles in a predisposed person, by preventing, as far as is in our power, the development of inflammatory affections of the bronchial or pulmonary organs. Such is the practical application of these new pathological views."

A few years later, in 1869, in my treatise on Pathology, I again spoke of the harmful influence of bronchio-pleuro-pulmonary irritation or inflammation, representing this as an occasional cause of pulmonary phthisis in speaking of the ætiology of the disease. I showed the importance and practical applications of my principles of pathology in the following statement, which is in my opinion the fundamental law of the therapeutics of the disease:—*The only substantial basis of prophylactic or curative treatment is furnished by the notion of imperfect nutrition and by a knowledge of the baneful influence of inflammatory processes.* Lastly, in my Clinical Lectures in 1872, I stated the same principles in analogous terms which I would recall *verbatim*, namely:

I. Caseation is at all ages the result of debility.

II. The origin of true tubercle is the result of debility.

III. The common forms of accidental irritation of every kind, affecting the larynx, bronchial tubes, or lungs, have a deleterious effect upon tuberculosis and phthisical lesions. This may happen in three ways. Firstly, in those who are healthy, but in whom predisposition exists, such irritation favours the development of tubercles or of the inflammatory changes which produce phthisis. Secondly, in those already affected it gives rise to a fresh development of tubercles. Thirdly, it aggravates and hastens the course of pre-existing disorders.

IV. Fever is a process of consumption.

"These four principles," I said at the end of my statement, "regulate the whole therapeutical treatment of phthisis, including as they do the reasons why certain plans of treatment should be excluded, and at the same time indicating and stating precisely to what extent medical supervision is advisable."

Since 1862 these ideas have widely extended; they were received with general approval, and the most eminent observers have considered their probability, facts which seem to show that they contain truth. The historical account of their progress, however, has not always been truly represented. In the enthusiasm of general agreement it was possible, as was done, to lose sight of the respective participation of each observer, and specially of those who were the first in date.

From the fact that these principles were universally admitted and applied, it was believed that this had doubtless been always the case, and that no reason existed for believing any phase of evolution to have occurred; thus, voluntarily or involuntarily, the first observers became completely forgotten.*

I could not approve of a method which, if carried to its extreme point, would suppress the historical consideration of every question, each writer considering it as merely dating from the moment when he takes up his pen. I desired therefore, and most legitimately I think, to recall my early intervention in this reform, connected as it is both with the pathology and treatment of pulmonary phthisis, and the principles of which regulate, as I have just said, all the therapeutical treatment of the disease.

This being said, I now commence the consideration of my subject, after warning all against the too widespread opinion that pulmonary phthisis is an incurable disease.

It seems to me that, considering the conclusions recently derived from pathological histology, it would be of the greatest

---

* Thus, for example, Lebert, in his book written in 1879, not only upholds the same pathological principles, but also borrows the designations which I composed, such as "phymatogenous irritation," "phymatogenous diathesis," "constitutional hypotrophy," and "dystrophy," without caring to mention the name of their author.—NOTE OF AUTHOR.

benefit, while this again would be a most advantageous opportunity, to proclaim more earnestly than ever the curability of this disease, and therefore the necessity of an active plan of treatment.

As already said, the microscope has not failed to show the anatomical unity of pulmonary phthisis. The memorable works of Grancher * (1872–1877), Charcot,† and Thaon ‡ have proved, that to a less extent than and independently of simple inflammatory products, the same elements, with the same arrangement and evolution, are found in the masses ascribed to caseous pneumonia as in the typical granulation. There is scarcely any difference between them except in size; the tubercle in the one case being small or miliary, in the other large and massive. The duality is therefore only in appearance, and whether the lesion be small and in isolated nodules, or large and composed of homogeneous masses fused together, it is found microscopically to consist in all cases of real tubercle. Phthisis, in fact, is invariably tubercular, and the anatomical unity of this disease is established by histological analysis. The authority of the observers quoted is such that it scarcely seems needful to say that their opinion should carry conviction with it; for, though the domain of histology is constantly advancing, is it possible to doubt a statement which bears the signature of Grancher and of Charcot? In generalizing the conclusion, it is right perhaps to admit that more light may be thrown upon the question by the knowledge of future times; but with this reservation I, for my part,

---

* J. Grancher, a French physician, the subject of whose thesis for the Paris degree in 1873 was "De l'Unité de la Phthisie," and who wrote an article in the *Archives de Physiologie*, 1872, upon the same subject.

† J. M. Charcot, the eminent physician of the Hospital Salpetrière, Paris, wrote in 1860 a thesis entitled "De la Pneumonie Chronique."

‡ Louis Albert Thaon is also a French physician, whose thesis for the Paris degree in 1873 was entitled "Recherches sur l'Anatomie Pathologique de la Tuberculose."

accept this statement in the same form as it was laid down by the above-mentioned authors.

Now, the readmission of this unity, which must still be acknowledged notwithstanding the powerful attacks made upon it, might lead to the idea of fatalism and the expectant plan of treatment which would result from the theoretical conceptions of Laennec. Such a deduction would be radically erroneous. The conclusion acknowledged to-day in favour of the unity of tubercle regards merely its anatomical structure, establishing upon a fresh basis the histological unity of the disease, while its clinical duality is not called in question. This duality I fully maintain, and whether its causation, and specially the hereditary tendency, its symptoms and course, or the prognosis be considered, the pneumonic or caseous form, whether microscopically tubercular or not, differs and always will differ from the ordinary form of tubercular phthisis. This duality being the guide of the practitioner, his constant effort should be to decide in each case which of these two morbid processes is the cause of the disease. I admit then, as I repeat, such anatomical unity, and I should do so with pleasure were it not for the fact that in this I differ from eminent colleagues whose legitimate authority no one respects more than I myself. While consequently agreeing with them that the pneumonic or caseous form of phthisis is, anatomically speaking, tubercular, I also maintain that it is tubercular in a special way, having its own causes, mode of invasion and evolution; that it has special prognostic characters remarkable for their unusually serious nature in the early stages of the complaint, but relatively favourable when the disease is more advanced. With regard to its clinical characters and medical treatment, I believe it to be twofold, and to establish this duality I have done as much as is in my power. Its anatomical unity, however, obliges us only to recognize different morbid appearances when duality admitted two distinct kinds of disease; and this doctrine being once expressed, I maintain that pneumonic phthisis has those characters which I assigned to it, subsequent observation continuing since 1870 to show me that this is the case. Knowledge recently acquired by means of the microscope diminishes in no way the value of my previous conclusions; far from it, such information confirms by many and new proofs the salutary notion of the curability of tubercle. It will not, I think, be difficult to prove this.

In the first place, it should be considered what, from this point of view, are the consequences of the doctrine of unity; it establishes, it proves the tubercular nature of the pneumonic and broncho-pneumonic lesions which belonged to caseous phthisis when the belief in duality was prevalent; this doctrine I not only accept, but adopt, since it is in my eyes of inestimable value as proof of the possible cure of tubercles in their most diffused form. It is certain, in fact, that these pneumonic processes, which were called but a short time ago caseous, are curable. Speaking only from personal observation, the cases analyzed in my Clinical Lectures, and those which I have since seen to the number of four, prevent my having any doubt on this point; as I have affirmed, I assert to-day, and always shall assert, even though meeting with no other examples, that the caseous forms of pneumonia which I have termed phthisigenic (*phthisiogènes* *) are exceptionally perhaps, but certainly curable. And since it is known to-day, by means of the microscope, that these forms of pneumonia have a tubercular nature, and constitute one of the most important forms of tuberculosis, it is evident that the new interpretation of histology fortunately enriches the balance on the side of the curability of tubercle, by transferring to pneumonic tuberculosis the favourable cases wrongly considered as simple caseation. I invite your special attention to

* *Phthisiogènes* (φθίσις, phthisis; γεννάω, I produce) means phthisis-producing.

this unforeseen and unmentioned consequence of the doctrine of unity.

Nor is this all; the histological works of which I have spoken, specially those of my excellent friend and eminent colleague Grancher, have established the importance and frequency of a healing process which can arrest the development of tubercle at any moment of its evolution, and whether of large or small size, and transform it into an innocuous product having thenceforth no action upon the organism or neighbouring parts. This process consists, as I have said, in fibrous transformation of the neoplasm; a process which is neither exceptional nor absolutely rare. Allow me to recall the important declaration of Grancher on this point: "That which (without considering its anatomical characters) distinguishes the evolution of tubercle from that of cancer is the natural tendency of tubercle to become fibrous." "This, then, would be a transformation due to the character of the lesion, and not a fortuitous and irregular change."

Every form of tubercle, in fact, is from the first liable to two forms of transformation of opposite kinds—caseous evolution at the centre, and fibrous transformation at the periphery. It is upon which of these two forms of change preponderates that the ultimate destiny of the neoplasm depends, extending as it does and involving the tissue of the organ in its own destruction should caseation and softening occur, whilst it is stationary and without harmful influence on the neighbouring parts should fibrous evolution take place in the whole diseased part. The lesion is then cured.

This curative process is not limited to the primary development of tubercles; it is doubtless so much the more beneficial, so much the more likely to ensure the return of the affected organ to a healthy state, as the lesion is nearer its commencement, this fact implying also a less extensive change of tissue; at the same time, in even the most advanced stages, the disease may receive benefit from this process of repair. Thus at all times during the stage of softening, when the central part of the lesions is in full retrogressive softening, when secondary tubercular formations have already formed around the primary lesion, the zone farthest removed from the centre, which is termed embryonic * and has a specially phymatogenous character, may prevent the disease from extending beyond the protective barrier due to fibrous transformation. The condition of the parts included within this defensive circle may subsequently improve from the occurrence of absorption and elimination, but even should they remain in the same condition, the danger is now arrested. Owing to this limitation which may be termed providential, *the patient has now the power of living with tubercular lesions that can no longer be injurious.* So long as his constitutional condition, and the pathological disorders to which the respiratory tract is liable, leave the barrier unbroken, so long as other tubercular masses do not develop at a distance from the destroyed focus, he is free from the ordinary troubles of the complaint which affects him. He is cured; though the cure, I allow, is *relative*, and its persistence can only be insured by watchful and unremitting medical attention. It may be said, on the other hand, however, that the possibility of such a cure exists, not only at the outbreak of tuberculosis, but during the whole stage of softening.

One case should be mentioned, chosen from many analogous ones, which will convince you of the reality and lasting nature of such a relative cure.

It is now eight years since my eminent colleagues, the Professors Botkin and Rauchfuss (of St. Petersburg), honoured me by placing under my care a young Russian, aged

---

* It will be remembered that the cells at the centre of tubercle are withered and granular, whilst beyond the margins of the growth is a zone of connective tissue in which cellular proliferation occurs.

fifteen years, affected by tuberculosis, limited to the right lung. Considering the account of the first outbreak of the disease, which had the sudden origin and acute course of uncomplicated pneumonia, considering the results of stethoscopic investigation which I made five months after the outbreak of the first symptoms, there was no doubt as to how the tubercles were developed. This was a typical case of pneumonic tuberculosis, without tubercular antecedents in the family. The general condition of the patient was gravely compromised, and the marked emaciation, the almost invariable occurrence of fever at night, made the case appear in all respects one of confirmed phthisis. Still the investigation showed that all traces of pulmonary ulceration were absent, for whilst the left lung was in a healthy condition, the greater part of the upper lobe of the right lung was thus found to be in a state of consolidation beginning to soften, this region having been affected by the primary pneumonia. Dulness in front and behind, and over the whole of the same part a mixture of bronchial breathing during both times of respiration, subcrepitant *râles* produced by coughing, were signs indicating the remains of caseous, or, as it should now be called, tubercular pneumonia. Owing to the injury already done to the constitution of the patient, and the nocturnal fever, not only was it to be feared that the existing lesions would persist, but also that they would more or less rapidly extend. Under the influence of treatment principally directed to restore the general health, under the influence of energetic counter-irritant remedies replaced at seasonable times by antipyretics, the young man's constitution improved in a remarkable manner, his strength became greater, the fever ceased, and the feared encroachment of the disease did not occur. The first winter was spent in Paris. In the spring it was found that the patient had notably increased in weight, and that his general condition was most satisfactory. Though no local change whatever had occurred, an imminent danger had been averted, and the patient, as said just now, had been placed in such a condition that he could live, notwithstanding the tubercular lesions by which he was affected. The field was now open for the application of those therapeutic modes of treatment which it was previously necessary to defer. Aërotherapeutics (*aerotherapie*), residence in such climatic stations as answer rigorously to the indications presented, have completed the work began; during each year something has been gained, although on three different occasions we had to meet intercurrent attacks of bronchitis, always accompanied by fever and slight hæmoptysis. Now for two years no such accident has occurred. I saw the young man again at the end of 1879, who presented every appearance of being in good health; he has now grown more stout, and lives in the ordinary way, being cured of his disease, although but relatively so. The restoration of constitutional health has arrested the injurious influence of the diseased focus which still exists, but remains in a harmless and inert condition as far as physical change is concerned. Though less extensive, perhaps, there is, however, always dulness at the apex of the right lung; the superficial respiration is weak and indistinct, whilst the deep is undoubtedly rough and almost bronchial; the cough produces moist sounds which simple respiration cannot do. It is possible that more complete improvement of the local condition may occur at a later date; but even admitting that the condition of things remains as at present, the patient still has all the advantages of incomplete recovery, and the fact enables the importance of such a cure, which I have termed relative, and of which I have observed many other equally conclusive examples, to be duly recognized.

The transformation of such a relative cure into absolute recovery, the possibility of which I have just mentioned, is not a mere supposition. In some cases that special condition which is characterized by complete constitutional recovery and the persistence of local lesions is but the first step in the work of recovery. After a variable time a more favourable stage ensues, signalized by the absence of appreciable local change. I have already observed in two cases such a total recovery. The first was in a young Swiss, of whom I shall speak at greater length when discussing the climatic form of treatment; the second in a young man whose history was given in my lectures at the Lariboisière Hospital. When these lectures were published (1872)—that is, nine years ago—this patient had only obtained the relative cure of which I have spoken, but during the two years which followed, continual improvement took place in the condition of the lungs. Since the end of 1874, no worse symptom has been discovered than a notable weakness, or perhaps abnormal roughness of the respiratory murmur at disseminated points, corresponding to the position of the primary lesions. Since then I have examined the young man frequently during each year, and without ever detecting any unfavourable symptom, the recovery remaining complete and absolute. His case again was one of pneumonic tuberculosis with an acute onset, which I called at that time, owing to the prevalence of the dualist doctrine, phthisigenic pneumonia (*pneumonie phthisiogène*); but the process of cure by fibrous transformation, which has just been mentioned, is not limited, as I should at once say, to this particular form of phthisis, but may be equally observed in granular tuberculosis with gradual onset, that is to say, in the ordinary form of the disease.

Curable as tuberculosis is by the fibroplastic process whilst in the crude state, and during the whole period of softening, it is so also when ulceration occurs and caverns form. The knowledge that recovery is possible under these circumstances dates from ancient times; the facts which establish both the truth of this statement and the anatomical mechanism by which it occurs are classical, and there is no need to insist now upon these points. I shall confine myself to reminding you of the different aspects borne by the healed cavern, as shown on numerous occasions by pathological anatomists. The principal appearances borne by caverns are of the number of four: the cavity persists being empty and communicating with the bronchial tubes (fistulous cicatrix [*cicatrice fistuleuse*] of Laennec), while the parts surrounding it are consolidated, infiltrated with pigment, and puckered by shrinking; or the cavity is filled with calcified tubercular matter; or occupied by a fibro-cartilaginous mass due to vegetation from the connective tissue in the wall of the cavity; or it disappears owing to the opposed surfaces becoming glued together, and forming a linear cicatrix of variable thickness, and of fibrous consistence where the bronchial tubes terminate, ending in a *cul de sac*. The surrounding lung becomes emphysematous in compensation for this; the pleura is thickened and puckered, and the wall of the chest is depressed unless dilatations of the bronchial tubes fill the empty space. At the centre of the cicatricial bridle may be found the remains of what the cavern contained in the form of chalky pulp, or solid calcareous concretions. In all cases the healed cavern is surrounded by a zone of interstitial pneumonia (*sclerosis*), which at the commencement of the healing process circumscribes and encysts the loss of substance, and prevents its injurious influence upon the neighbouring parts. Molecular disintegration and elimination of the contents do the rest. Observe, then, that one finds here again that fibrous formation is the principal agent in the cure, though it is no longer the tubercle which undergoes this healthful evolution, but the surrounding tissue. It is this which is transformed by the sclerous neoplasia (*neoplasie* *) into a preservative enclosure, and prevents the irritating and encroaching action of the ulceration from spreading to the healthy parts.

* *Neoplasie* (νέος, new; πλάσσειν, to form) means the formation of a new growth, reference being here made to that of fibrous tissue.

These conditions of the lung which testify so plainly to the curability of phthisis at the period of excavation are not absolutely rare; I have often seen them after death, and in the two last years have had the opportunity of observing two cases, one showing recovery with persistence of the cavern, the other with total occlusion of the ulceration. The first of these cases was that of a woman, aged sixty-five years, in whom it was easy to recognize the existence of a large dry cavern at the apex of the right lung; there was, however, no symptom of actual phthisis, and the answers of the patient showed that all coughing and expectoration had ceased for many years, the exact number of which she could not tell, but believed to be more than twenty. She was admitted into the hospital for disease of the heart, to the progress of which she succumbed. The autopsy confirmed the diagnosis of cured phthisis. In the upper lobe of the right lung was found a cavern exceeding a large walnut in size, invested in its whole circumference by a thick fibro-connective membrane, and kept open by adhesion of the whole lobe to the thoracic wall. There was no other trace of tubercular lesion. The primary focus had existed alone, the diathesis being, so to speak, exhausted by this primary manifestation, and the local improvement by cicatrization of the ulcer had been the cause of the patient's recovery. In this as in all similar cases, the early loss of substance, the formation of a cavern by softening, the destruction and elimination of the diseased tissue, had evidently been a favourable event. This possibility will be again considered.

The second fact is no less remarkable. The patient, a woman of fifty years, succumbed rapidly to tubercular pneumonia, hæmorrhagic in character, and affecting the central and lower part of the left lung. At the same time that the autopsy verified the existence of this recent affection, we found that the upper lobe of the same lung was completely transformed into a compact, homogeneous, fibro-cartilaginous mass, whose continuity was only interrupted by dilated bronchial tubes, the walls of which were firmly indurated, at many points infiltrated with calcareous matter, and when their canals were opened by such incisions as the examination required, the collisions of these chalky masses produced in the hand which compressed the altered lobe, the exact sensation of a sack of hazel-nuts. Had we in this case tubercular pneumonia of the apex, or a simple granular product? I cannot say; but this is certain, that the upper part of this lung had been long riddled with caverns of small size, that the cicatricial closure of these ulcerations had brought about cohesion of the fibrous products, that the connective tissue by retraction from the bronchial tubes had caused their dilatation, and that, lastly, the curative fibroplastic process acting in excess had reached the parts which were free from ulceration, and had thus brought about the total transformation of the upper lobe into an inert mass of fibro-connective and fibro-cartilaginous tissue, without vestige of pulmonary tissue.

The local cure was far more complete and more firmly established than in the preceding case. To this recovery the patient owed complete immunity from disease during a large number of years, as shown by the anatomical characters presented. The diathesis, however, was not exhausted, or at any rate reappeared when the woman was in bad hygienic conditions during the last three years of her life. Its very acute manifestation at that time quickly effaced every sign of recovery from the first attack. The probable date of the old attack of tuberculosis was totally unknown to me, and this for reasons which I would now mention. In the first place, when the patient came into my service, where, as I should mention, she remained but three days, her condition was so serious that a complete history could not be obtained; and secondly, even had it been otherwise, I should not perhaps have given to my questions the right direction; for I found over that part of the upper lobe which was transformed into a fibrous mass, absolute dulness, tubular breathing, and marked bronchophony. I found the same signs over the middle and lower parts of the lung, with this difference alone, that the tubular breathing was at that part conjoined with moist *râles*, which was not the case above. Thus I believed in the existence of a similar lesion of almost similar intensity in all parts of the lung. The autopsy first revealed to me that the pneumonia did not affect the upper lobe (this being due to the fact that no pulmonary tissue remained at that part), and that the stethoscopic phenomena perceived in this region were not due to a lesion actually in evolution, but to the indelible remains of an old alteration cured.

It should be remarked that, notwithstanding the fatal termination of the second attack, in this as in the former case the formation of cavities was a healthful process. This change was the precursor of cicatrization, which could never have been completely effected without previous removal of the pathological products.

I would urge upon all the fact that recovery from phthisis is possible at the stage of ulceration, as proved not only by post-mortem examination, but by clinical observation. This may fairly be allowed to show the same truth when the facts are clear, when a consecutive history of the disease can be ascertained with exactness, and when it is possible to compare rigorously the present state, considered as one of recovery, with the morbid condition which existed before, as ascertained either by direct observation, or from a report written by physicians, whose honesty and competence are to be trusted. I have seen a certain number of such cases which in my opinion fulfil all the requisite conditions. The most remarkable was perhaps that of the Duke of R——, and although it is already mentioned in my Clinical Lectures, I have no hesitation in recalling the principal particulars, so anxious am

I to impress upon every mind the valuable idea that phthisis is curable even when caverns have already formed.

The Duke of R—— then, who was under my care for gastric troubles in 1870, observed one day, "I, whom you now see, have been cured of pulmonary phthisis." I at once asked for permission to auscultate the chest, and making this examination with extreme care, I recognized that the left lung was perfectly healthy, whilst at the apex of the right lung I found posteriorly, both at the inner part of the supra-spinous and the upper part of the infra-spinous fossa, undoubted signs of consolidation. There, though the dulness was not extreme, its existence was clear, the respiration was markedly bronchial, and there was bronchophony at these points; no *râle*, however, was perceptible, even when the patient coughed, and the consolidation was undoubtedly dense, compact, and homogeneous. Upon recalling the symptoms presented by the patient whom I previously mentioned, a striking resemblance is found to exist between the two cases. The patient observed to me at the close of this examination, "At the very point, where you have been listening for so long, there was a cavern." He then informed me that a consultation had taken place between Chomel * and Louis † which leaves no doubt on this point. Fifteen years before, it had been recognized and asserted that the Duke of R—— was affected by phthisis, and he then presented, besides other troubles, a cavern on the right side. The evolution of symptoms had been as follows :—The patient had coughed for some months, and had several attacks of hæmoptysis, when he was affected by an acute disease of the

* Auguste Francois Chomel, who died in 1858, was a French physician, first attached to the Hospital de la Charité, and subsequently to the Hôtel Dieu at Paris. He succeeded Laennec as clinical professor at the École de Médicine in that town, and wrote "Pathologie Générale" and other works.

Pierre Charles Alexandre Louis, who died in 1872, was honorary professor of the hospitals at Paris. He wrote "Recherches Anatomiques, Pathologiques et Therapeutiques sur la Phthisie," "Recherches sur la Fièvre Typhoide," and other well-known books.

chest—in all probability broncho-pneumonia—during which complaint hæmoptysis occurred for the last time. This acute attack gave place, at the end of a few weeks, to a chronic condition during which all the phenomena of consumption successively showed themselves. The prognosis of the physicians had been most unfavourable; upon which the patient, refusing all medicine, recommenced taking wine and cordials with a tonic diet, and departed for the waters of Panticosa,* a nitrogenous and saline spring in the neighbourhood of Cauterets.† In this favourable climate he gradually improved, until in less than a year he had completely recovered. It may be added that the cure was final, since when the patient consulted me after an interval of fifteen years nothing abnormal was found, except an indurated patch in the upper lobe of the right lung. In such conditions clinical demonstration seems to be as clear as post-mortem examination.

I have repeatedly stated that, with respect to curability of the disease, the formation of caverns is in some cases a favourable occurrence; and this I again affirm to be true, though contrary to the opinion which is generally held. This opinion is based upon an error of interpretation which should be mentioned. The course of pulmonary phthisis has been

* Panticosa is a place in Spain (Arragon), on the southern aspect of the Pyrenees. See "Dictionary of Watering-Places" (L. Upcott Gill), part ii.; "Guide to Spain and Portugal," Henry O'Shea; "Handbook of Spain," Richard Ford (J. Murray); "Health Resorts," J. Burney Yeo, M.D.; "Curative Effects of Baths and Waters," J. Braun; "The Baths and Wells of Europe," J. Macpherson, M.D.; "Guide to the South of France," C. B. Black; etc.

† Cauterets is a town in France, situated in the arrondissement d'Argeles in the Hautes Pyrenées. See "Dictionary of Watering-Places" (L. Upcott Gill), part ii.; "Health Resorts for Tropical Invalids," Moore; "Handbook for France" (J. Murray), part i.; "Health Resorts," J. Burney Yeo, M.D.; "Curative Effects of Baths and Waters," Julius Braun; "The Baths and Wells of Europe," J. Macpherson, M.D.; "The Principal Baths of France, Switzerland, and Savoy," E. Lee, M.D.; "The Mineral Waters of France," A. Vintras; "Dictionnaire Encyclopédique des Sciences Médicales, Paris;" "Curative Influence of the Climate of Pau," A. Taylor, M.D.; "European Guide-book for English-speaking Travellers," Appleton; etc.

supposed to be continual and always to progress with similar rapidity; consequently the ulceration producing caverns has been looked upon as the last, the final, and consequently the most serious of all lesions in the disease. None of the statements in this proposition is applicable to every case; the course of phthisis is not always continual and progressive, as it may advance by means of attacks often separated by long intervals of time. As far as each tubercular focus is concerned, the cavern is certainly the final lesion from a chronological point of view, but it is not on that account the final lesion of the disease; still less is it the most important if its severity be considered, nor does it indicate greater peril, or approaching danger.

Considered by itself and in itself alone, a cavern causes no constant modification in the prognosis; far from rendering it more gloomy, this may be the signal for legitimate hope, everything depending upon the condition of the patient at the moment when the ulceration occurs, upon the extent of the latter, and the importance of pre-existing lesions. What, in fact, is a cavern? Nothing more than ulceration, the loss of substance resulting from the destruction and elimination of the tubercular tissues. If the diseased focus is of any size, such ulceration is the preliminary condition needful for its cure. When, then, in a certain tubercular focus such a condition is realized at an early date, when the general condition of the patient is not as yet too seriously compromised by the disease, and the walls of the cavity thus formed do not themselves contain tubercles, the elimination of tubercular elements, which are morbid products in themselves harmful and dangerous, permits the work of cicatrization to go on, a work which is so much the more certain and rapid as the constitution of the patient is less affected by the disease. It is therefore by no means true to say that the cavern, even at an early stage, indicates more severe or immediate danger. In the conditions which I have mentioned, it may be the signal and means of the most healthful repair. As long as the focus which is cured by means of elimination is solitary, the organism freed from its abnormal product owes to this fact the termination of its phymatogenous tendency, and not merely is a local lesion temporarily cured, but the disease itself terminates, either completely, or until some harmful hygienic or pathological process gives new activity to the slumbering tubercular diathesis.

Such a favourable evolution is not absolutely rare in pneumonic phthisis, in which the foci are minute in size and small in number. Already in my Clinical Lectures I have related two cases which are most conclusive, and since then I have seen two others in my practice. In one of these the recovery has occurred so recently that I cannot certify it to be final, but in the other it has already lasted three years, and without its completeness being impaired. In such conditions, I repeat, the early ulceration and elimination are phenomena of good augury, and the formation of a cavity may well be considered the final stage, not because it represents, as is commonly believed, the last and most serious epoch of the disease, but, on the contrary, from its signifying the origin of repair and cure.

It naturally follows, on the other hand, that if the tubercular foci are numerous and spread through both lungs, or if the elimination of one or more of these is accompanied by fresh tubercular development, which again gives to the exhausted organism the same work to perform, or lastly, if the ulceration occurs cotemporaneously at different points, it follows, I say, that the stage of excavation is the signal of decided aggravation, since on account of their extent or number, on account of the new troubles of which they are the direct cause, the ulcerations, whether their surface be clean or foul, cannot be made to cicatrize by the exhausted organism. Deferring for an instant to consider the special nature of the lesion, I feel sure that all will be struck with the justice of this distinction. The condition of things, considered as a whole, is in fact the same as in all diseases presenting foci, whose cure requires first elimination, and then cicatrization of the loss of substance due to this elimination. Should the foci be of minute size, circumscribed and in small number, and the general condition of the patient be good, the work of repair may be effected, and elimination is then the first stage of recovery. Should the local and general conditions be of an opposite character, the work of cure is more than the organism has power to accomplish, and elimination is the first sign of irremediable aggravation.

Thus it is that ulceration of the lung may occur at an early or late stage of the disease, may be favourable or injurious, and has of itself no meaning, since this depends upon the general condition of the patient. It is a simple incident without certain connection either with the period or the prognosis of the disease. I attach great importance to the personal ideas which have just been mentioned, since they render the prognostic appreciation of the disease more clear, and may prevent an unfortunate mistake in the treatment. What, in fact, happens, should the physician consider the formation of caverns as certain evidence of a most severe final stage? In recognizing their formation, he at once modifies his opinion as to the subsequent course of the disease; in this he may be mistaken, and be subsequently undeceived by what really occurs to the great advantage of the patient. Besides this, he thenceforth looks upon treatment as useless, and being discouraged, gives up all therapeutic intervention, having prematurely by a mistaken prognosis pronounced against it the sentence of impotence.

To sum up what has been stated, pulmonary phthisis is curable in all its stages. This is the prolific notion which presides over the whole history of the disease, and which should unceasingly inspire and direct all medical action. The incurability proclaimed by Laennec and his immediate successors is disproved by pathological anatomy and clinical observation. None should therefore allow themselves to be influenced by such condemnation, which is but a historical souvenir. When the existence of tubercles in the lung is recognized, it should not be inferred from that moment that he who has them is doomed to death in consequence of their presence. Should it be found that the tubercles soften and a cavern forms, it should not be believed on this account that all is lost. It has been shown that this is not the case, and the natural tendency which tubercle has to fibrous transformation, that is to recovery, should never be forgotten. Before being discouraged, the physician should search and examine incessantly whether the patient is in the requisite conditions for such favourable evolution to occur. If all hope of absolute recovery must be abandoned, a relative cure should be sought, and every exertion be made to place the patient in such conditions that he can live notwithstanding lesions which are now irreparable; in a word, the plan adopted should be to strive and strive always with the unshaken confidence which may be drawn from the notion that recovery is possible. The enemy *can* be conquered; this is the idea which should engender and sustain every effort. It is certain that this conviction is the first condition of success, since it is absence of faith in the possibility of cure which prevents the adoption of all therapeutic treatment.

CHAPTER II.

CONDITIONS WHICH INFLUENCE THE CURABILITY OF PULMONARY PHTHISIS.

Influence of age and extent of lesions—Conclusions derived from the ætiology of the disease.
Hereditary phthisis—Innate phthisis—Acquired phthisis—Primary phthisis, acquired at an early or late age—Secondary phthisis—Scrofulous phthisis—Arthritic phthisis—Diabetic phthisis—Herpetic phthisis.
Data furnished by anatomical appearances—Ordinary form—Pneumonic form—Characters special to each form as regards its prognosis and curability—Necessity of diagnosis with respect to pneumonic phthisis.

THE curability of pulmonary phthisis has now been established at every stage, and in its two forms, the ordinary and the pneumonic. This general conclusion would, notwithstanding its engrossing interest, be almost fruitless were it not to be completed by an investigation of those conditions which may contribute in some way to this happy result. The favourable indications are by no means the same in different cases, and it is of importance, both as regards the method and continuance of the treatment, that the physician should be well informed as to the circumstances of different kinds upon which the possibility of absolute or relative cure depends. To have this knowledge numerous details must be considered, details which have not by any means the same importance, as the following analytical review and discussion will show to be the case.

In the first place, whatever form of the complaint, whichever patient be considered, the chance of recovery is inversely as the duration of the disease and extent of the lesions. It is easy to understand why these conditions have so much influence. The longer the disease has lasted, the more the constitution is affected, and it is the extent of this change which decides the whole question of recovery. On the other hand, when the general health of the patient is so good that recovery seems possible, it is still the more difficult and uncertain, in proportion as the lesions needing cure are of larger size and more numerous. This is evident, and it is unnecessary to consider such truths at greater length in order to prove their importance; on the other hand, it may be added that they themselves show the importance of early medical intervention, since the duration of the disease and the extent of the lesions are in most cases directly connected as they advance in parallel lines. In consequence of this, the early treatment of the disease has the double advantage of acting upon a constitution which is less affected by it, and upon lesions which are of smaller size and more recent formation.

The general conditions being thus recognized, the indications furnished by ætiology as to the question of recovery will now be considered.

It will be remembered that with regard to its origin three varieties of phthisis may be distinguished—the hereditary, the innate, and the acquired form.

Of these the hereditary form, from the mere fact of its being hereditary, offers the least prospect of recovery. This is for two reasons. In the first place the diathesis, that is to say the unfortunate disposition of the organism to form tubercle, has in these cases its greatest power; from the beginning of fœtal existence this forms part of its physical individuality; as one of its attributes it remains in an active condition so long as life remains, and without being exhausted by the primary manifestations of its existence. Thus, if the first indications are cured, others almost undoubtedly follow, fresh labour being imposed upon the organism to which it must eventually yield. This peculiar power of the inherited diathesis is so much the greater and more menacing that the ancestors and relations have suffered in greater number, and when both parents are affected by phthisis the activity of the original tendency is irresistible. Nor is this all. Chronic phthisis, of which alone we have been speaking, when hereditary, almost invariably takes the ordinary anatomical, that is the miliary form of the disease; and, considering all the facts which I have observed during the last twenty years, I would assert to-day, as I have already done in my previous writings, that this is less frequently cured than the pneumonic form, should the latter become chronic. This point will shortly be reconsidered.

If, however, hereditary phthisis is less curable than other forms of the disease, it may receive much benefit from preventive treatment, and offers in my opinion specially favourable conditions for such prophylactic care. The fact of the disease being inherited reveals from an early age the impending danger. Whether the patient is affected early or late in life, the disease is at no time an unforeseen surprise, but the realization of what has been long dreaded, and thus a prophylactic course of treatment may be set on foot much earlier than in other forms of the disease. It may precede by many years the first indication that the virtual is being transformed into the real disease, and owes to this long continuance the greater probability of its being effective. Perceiving from the first any approach of danger, the physician has time to act, and this character of the hereditary disease with respect to preventive treatment compensates to some extent for its known incurability. In a family which inherited phthisis from one parent, and which had had the misfortune to lose two infants from tubercular affections, by means of energetic treatment continued during many years, I have succeeded in modifying completely the constitution of the three other children. Having passed the age at which the others were affected, there is not the least sign of disease, and I am convinced that the improved constitution which they owe to therapeutic treatment definitely protects them from the complaint by which they are threatened. I have observed two other cases of the same kind, which, however, were of less value in demonstrating the above facts, owing to occurrence of the illness in families in which the hereditary influence was of a collateral kind.

Innate phthisis, which must not be confused, as I have already said, with the hereditary form, is observed in the descendants of those who, though not tubercular, are weakened by scrofula, cachectic diabetes, alcoholism, or simply by bad hygienic conditions; besides these causes the innate form may also be due to consanguineous marriages. The phymatogenous* diathesis (*diathèse phymatogene*) exists in such children from their birth as in the preceding form; being, however, in this case innate, and not hereditary, since the parents were not affected by it. Should it not be recognized that the diathesis is innate, the persons affected by this form of phthisis are naturally believed to have acquired the disease, and the number of such cases is therefore greatly exaggerated. It should be remarked, as already stated in my treatise on Pathology, that the bad conditions in the parent or ancestor which explain the innate character of the diathesis, are precisely those which produce the acquired form sooner or later after birth. The only difference is that in one case two generations are required by the tendency which exists to produce tubercle, whilst in the other one lifetime suffices that this may be done. From such a point of view it may be truly said that tuberculosis is the common result of all forms of constitutional deterioration either in the family or in the individual.

* Phymatogenous means "tubercle-producing," as explained in the note of chap. i. p. 7.

Thus, created and characterized by elements which belong, as far as their cause is concerned, to the preceding generation, to the injurious influence of which the product passively submits, the innate diathesis is to some extent less incompatible than the hereditary with the idea of recovery; this, in my opinion, is from its occurring less often in the miliary form, and from its being often connected with scrofula, so as to share the good or bad prospects of scrofulous phthisis, of which we shall soon speak. Thus in the innate form there is at least a possibility, a chance which removes from it the character of absolute incurability, which we were bound to admit in hereditary phthisis, when this diathesis was once realized. Otherwise prophylactic treatment has in both cases the same efficacy, since the characters of the innate diathesis are significative enough to call attention to the disease early in its existence, and to reveal to the attentive physician the necessity for active intervention.

The forms of acquired phthisis which I distinguish are, as is well known, the following:—

In the first group I place those cases in which the pulmonary tuberculosis, being spontaneous and independent of other diseases, could only be due to general debility, to that insufficient or improper nutrition which is the basis of all forms of phthisis; such is "the primary form of acquired" phthisis. In the second group I place those cases in which the pulmonary disease is connected with a constitutional affection either present or past, and to the existence of which it may be rationally imputed; this is "the secondary form of acquired" phthisis.

With respect to curability, primarily acquired or idiopathic phthisis is the most favourable of all its varieties, whether the anatomical form be granular or pneumonic, and if the extent of the lesions be supposed the same in all cases. One condition alone will have produced it, namely, constitutional malnutrition, resulting in weakness which itself, or owing to the effect of ordinary irritation (an occasional cause), gives rise to the formation of such a low product as tubercle. This weakness is the consequence of such physiological or pathological conditions as result in excess of organic loss or insufficiency of nutrition. It is accidentally acquired, not belonging to the constitution of the patient, and on account of its accidental character may be and is most efficiently resisted by treatment which is practised at a sufficiently early date. Judged, then, with regard to constitutional repair, the condition is better than in the preceding forms; and since, again, the pulmonary lesions in the ordinary kind of acquired phthisis develop gradually, and remain for a long time circumscribed in size and small in extent, one perceives that the affected part offers most favourable conditions for recovery, whether the general state of the patient be considered, or his chance of recovery be supposed to depend upon the condition of the local affection.

To have an exact and complete idea of the question of cure in the primary form of acquired phthisis, a distinction should also be made between phthisis acquired early in life, that is between the ages of fifteen and thirty-five, and that which occurs at a later age, namely, from thirty-five to forty, or at a subsequent date. When late phthisis is really primary according to my definition, that is independently of any existing or previous constitutional disease, it may be controlled more decidedly than any other form of the complaint by appropriate treatment. Observation shows that the ordinary causes of such tuberculosis are repeated chills in those who are depressed by excess of work, and poor conditions of life. In these cases the course of the disease is naturally slow, and constitutional repair is rapidly obtained by means of suitable treatment, that is by a total hygienic change. In soil thus happily modified the lesions may diminish, or at any rate may cease to increase, so that though absolute recovery may not follow, it is possible to obtain in many cases a relative cure, the importance of which can be now appreciated. Such late phthisis, owing to the age at which it occurs and the causes which produce it, is less influenced by diathesis than any other form of the disease, and this is the true reason of its being more easily cured.

The secondary forms of acquired phthisis, as already said, are, so far as their origin is concerned, due to and caused by present or previous constitutional disease. Scrofulous phthisis, which some observers, believing in the identity of the two affections, have regarded as the latest and most pronounced manifestation of scrofula, is the most frequent, and at the same time the least curable of such secondary forms of the disease.

With regard to curability, I cannot regard it so favourably as the primary form of acquired phthisis, though it is incontestably less formidable than the hereditary or innate form so long as the lesions are confined to the lungs. Its evolution is remarkably gradual, and this natural slowness in its progress is seen even in the pneumonic form, which never presents the rapidity and irresistible acuteness which too often characterize the other ætiological varieties. The physician has time to act, and scrofulous malnutrition, which is the basis of this morbid condition, whatever local affections it may have produced, is of such a kind as to be most surely controlled by therapeutic treatment. In this form, again, cure is therefore possible, and though complete recovery may not be attainable, a relative cure should always be expected and desired which may ensure immunity from the disease for many years, and sometimes for an indefinite period. The Duke of R——, and the young Russian whose history has been given, were both scrofulous; their form of phthisis was the scrofulous form; both have recovered, the former with complete cicatrization, the latter with the persistence of lesions in an inert condition, which for a long time have had no effect upon his general health.

On the other hand, when the lesions are not strictly confined to the lungs, scrofulous phthisis has no longer the favourable prospects of which I have spoken, and may be an extremely serious affection. This is because on the one hand glandular affections may expose the patient to the dangers of mesenteric, bronchial, or cervical adenopathy, and on the other because caseous foci which have not disappeared since the first development of scrofula in infancy may produce secondary infection resulting in acute miliary tuberculosis.

Arthritic phthisis is far more rare, and consequently of less importance than the preceding form. It may also be said that, except for convenience and rapidity of language, this term should scarcely be adopted. There is no proof that arthritis produces the disease, in the same way that scrofula is the direct cause of scrofulous phthisis, and the term "arthritic phthisis" should only be looked upon as an abbreviated expression of "phthisis developed in the arthritic." Besides this difference drawn from its ætiological cause, another fact plainly confirms the legitimate nature of my distinction. The truth of this is shown by the fact that the symptoms of pulmonary tuberculosis in the arthritic have certain characteristics owing to the nature of the region in which their evolution occurs. The disease appears late, and is developed slowly; the lesions are circumscribed and of small size; though occurring frequently, the attacks of hæmoptysis have not their usual gravity; the ulcerations when once formed are remarkable, often for their size, always for the comparatively slight effect which they have upon the general health; they are not ordinarily the seat of abundant secretion, but, on the other hand, become often desiccated, without producing expectoration to any notable degree. These particularities, however, whose interest I by no means disallow, are not such as by themselves to characterize the disease, and enable it to be recognized should the antecedents of the patient be totally unknown. It is from knowledge that the patient is arthritic that the idea of arthritic phthisis is formed, and the reverse clinical problem could not be solved and the character of the phthisis be understood from knowledge that the patient is arthritic. It may be remarked, that if the relation as to cause and effect between the arthritic disposition and tuberculosis is certain, the disease cannot be known in every case by the mode of its evolution, and it will be recognized that we are still far from knowing the fixed and definite morbid condition which exists in it, and that it is only on account of a distant and contestable analogy that this form of phthisis can be placed by the side of the scrofulous form. It is only with the benefit of this reservation that I intend to speak of arthritic phthisis.

The principal clinical characters of the so-called arthritic form of phthisis have but just been recalled; these characters lead one to suppose that the complaint is specially curable, and observation confirms this presumption. The slowness of its progress, which is but rarely interrupted by acute episodes, the extended duration of the period of consolidation, are circumstances which enable it to be hoped that the treatment may be effective, since they undoubtedly assure to it a long continuance. The primary lesion again, always circumscribed, may rest for an indefinite time in the state of consolidation on account of the fibrous transformation which precedes any work of softening; part of the upper lobe of one or both lungs is then lost as far as its functions are concerned, but the tuberculosis is cured even before the condition of phthisis has been reached. I believe that I have twice observed this eminently favourable evolution. Lastly, that most formidable aggravation, which results from the formation of successive tubercular deposits, occurs in but few cases, the lesions remaining confined to the parts which were affected by the first manifestation of the disease. Softening and elimination may occur without producing irretrievable destruction, and this stormy period may lead, with or without persistence of a cavity, to a more durable, and at times to a definite recovery. I saw a good example of this in a lady who for many years was under my care, and who now enjoys good health, though she had an empty and desiccated cavern at the apex of one lung for three years.

A more serious form of the complaint, diabetic phthisis, will now be considered, and specially as regards the question of recovery. I have had many opportunities of observing this disease, but have not as yet seen one case in which the prognosis could be looked upon as favourable. Theoretically, one could conceive that it might be otherwise if the wasting occurred early in the course of the diabetes, while that disease could still be arrested, and before the general condition was seriously affected; observation, however, shows that these conditions, relatively so favourable, are never fulfilled. Phthisis is a late complication in *diabetes mellitus*, occurring constantly when the patient is reduced by the illness, that is to say in the most serious form and at the most serious time of the disease, and it thus only precipitates the fatal termination, by adding to the dangers properly called diabetic, those of a complication which the now modified state of the patient renders totally incurable.

Finally, the herpetic form of phthisis should be mentioned. In my opinion, its existence is even more uncertain than that of the arthritic form, and can only be allowed in the following circumstances. A person has presented for some years the cutaneous manifestations of herpes, when, spontaneously or from the influence of some accidental or therapeutic cause, these manifestations end; that is to say, cease to occur during a length of time exceeding the interval which previously separated them. This cessation of the cutaneous symptoms is followed after the same length of time by the development of the first pulmonary disorder. Upon such an occasion the existence of herpetic tuberculosis or phthisis may certainly be suspected, but in all other cases this is an arbitrary designation merely expressing a supposition drawn from theoretical considerations.

When such conditions exist as those I have just named, the condition is, in my opinion, favourable as regards recovery from the pulmonary disease, so long as the indication is followed at a sufficiently early date, and the treatment directed accordingly. This assertion is made in a somewhat doubtful form, not because my convictions are uncertain, but because I have as yet seen only one case of the kind. It was in a lady aged forty, of a vigorous constitution, who, owing to arsenical treatment, and being twice cured of her complaint by the waters of Loeche,* recovered from catarrh limited to the apex of the right lung; this catarrh had become established slowly and imperceptibly, without notable alteration in the general health, and after the definite cessation of a cutaneous eruption to which she had been subject since the age of about twenty-five years.

Such, then, are the conclusions drawn from the ætiology of the disease with regard to the curability of pulmonary phthisis. This consideration, which is the basis of all treatment, will still be continued, and what opinion may be deduced from the anatomical form of the complaint will now be discussed.

Acute miliary tuberculosis being for the moment disregarded, inasmuch as it does not produce the clinical symptoms of phthisis, two forms alone remain to be considered, namely,

---

* Loeche, or Leuk, is situated in the Valais (Switzerland), at the junction of the Rhone and Dala. See "Dictionary of Watering-Places" (L. Upcott Gill), part ii.; "Curative Effects of Baths and Waters," Julius Braun; "The Baths and Wells of Europe," J. Macpherson, M.D.; "Switzerland and the Adjacent Parts," K. Baedeker; "Handbook for Travellers in Switzerland" (J. Murray); "Dictionnaire Encyclopédique des Sciences Médicales, Paris;" "The J. E. M. Guide to Switzerland," J. E. Muddock; "Hachette's Diamond Guide to Switzerland," Adolphe and Paul Joanne; "European Guide-book for English-speaking Travellers," Appleton; etc.

chronic miliary tuberculosis, or the *ordinary form* of the disease, and the *pneumonic form.*

Notwithstanding its more gradual development, and that its progress is at first less rapid, the miliary offers in typical cases a condition which is decidedly less favourable to recovery than the pneumonic form. This statement, already made in my Clinical Lectures, is deduced solely from the results of my own observation. It might be supposed in consequence that it was the effect of a casual sequence of cases; this opinion, however, can scarcely be held, since the sequence would have to be of too great length. The facts which I have studied since the publication of my lectures confirm the assertion, while I also find, upon comparing these two forms, sufficient reasons for the difference which they present as to curability. These reasons are certainly of an indirect kind, and not connected with the special form of the lesion; but, as far as the final and practical conclusion is concerned, this is, judging from all that I have seen, such as was just asserted. We will now survey the indirect or direct causes of the difference which exists between the two principal anatomical forms of tuberculosis, and this consideration will remove all that might appear paradoxical in my assertion.

The "miliary" is connected, so to speak, with the highest form of the diathesis; the hereditary and innate diseases are usually expressed in it, and this would be the cause of its greater resistance to the healing process. In this form the lesions are almost invariably bilateral, either from the first, or before the disease has been long in existence, which is another unfavourable circumstance owing to the influence exercised by the large extent of the lesions. In this form the lesions are more disposed to increase by successive miliary development as usually occurs in the disease, while one rarely observes that tendency which, on the contrary, is so frequent in the pneumonic form—I mean, to the formation of foci of great or small size, whose evolution may be favourable or unfavourable, but which remain the only lesion, either permanently or for a great length of time. In the miliary form the lesions are but rarely confined to the lungs; sooner or later similar changes occur in the larnyx, the intestine, and at times also in the brain, and thence additional groups of symptoms are presented, which aggravate the condition of the patient, either owing to their existence, or because they impede the adoption of useful treatment.

It will now be understood how numerous and important are the causes which diminish the chance of recovery in the ordinary form of miliary phthisis. At the same time, such unfavourable conditions do not always exist; this form may be connected with the acquired disease, the lesions in this case being also circumscribed, of small size, and confined to the lungs, the condition being then not only as good, but even better than in the pneumonic form, on account of the more gradual development of the primary lesions. The chance of recovery, in fact, is then entirely dependent upon the constitutional condition of the patient, which, as will be seen, overrules at all times the whole question of prognosis.

Before saying more, I would again recall to your notice, on account of its importance, a fact which I have already stated. The reason that the usual miliary form of the disease is less curable than other kinds, is neither the character of the lesion nor the course which it takes, the slow progress of which is, on the contrary, a favourable circumstance, but rather the usual ætiology which it has, the bilateral character of the lesions early in the disease from the successive production of new foci, and the frequent existence of serious complications in other organs. The presence of these unfavourable conditions is usual in the miliary disease, while, on the contrary, they are most often absent in the pneumonic form, which maintains for some length of time, or at any rate seems to maintain, the character of a local and accidental affection. This difference is in my opinion the special cause why ordinary miliary phthisis is less curable than the other form. When, however, as in exceptional cases, this disease presents the same conditions as the pneumonic form, it is quite as capable, when equally developed, of being arrested and cured, since it does not present at its outbreak the imminent and serious danger which characterizes the initial period of pneumonic tuberculosis.

The discussion which precedes enables me to pass rapidly over the causes which, from a general point of view, render pneumonic phthisis more specially curable, and which need but be briefly recapitulated. This form of phthisis is most often an acquired and primary disease, or perhaps is connected with scrofula, receiving then the virtual advantage specially conferred by this ætiological cause; the lesions are either confined to the lungs, or affect other organs at a late period in the complaint, and most often unilateral they remain for a long time, and perhaps definitely of the same size as at the outbreak of the disease. These circumstances are favourable in character, but obviously can only exist when the pneumonic tuberculosis becomes chronic. The primary period of the disease is attended with constant danger; the first outbreak may prove fatal without its acute character having been once interrupted, so that this complaint may constitute a form of phthisis (galloping phthisis) no less serious than acute miliary tuberculosis, to which, in clinical history, the diffused or pneumonic tubercle may exactly correspond. We shall return to the history of its progress when considering the treatment of this form of the disease. I repeat, however, that when the illness continues so as to become chronic, constituting a disease to which the name of phthisis may be more properly applied, and when the conditions already mentioned are exactly realized, this form is that which, from the character of its subsequent progress, presents in my opinion the best prospect of recovery; in these relatively favourable conditions there are but the general health of the patient to consider, and the existence of secondary lesions, either pneumonic or miliary, which are unfortunately but too frequent, and which in a very short space of time may altogether modify the evolution of the disease. This explanation will justify the peculiar prognosis which I assigned previously to pneumonic tuberculosis, a prognosis which has this special character, that at first it is more unfavourable than in any other form of the complaint, miliary granulosis being alone excepted, while during the time which follows the acute outbreak it is the least serious of all the different varieties.

Most cases of recovery, whether absolute or relative, which I have observed (and of which striking examples have been already mentioned) belong to the pneumonic form of tuberculosis, and this fact in my opinion strongly confirms the ideas which I have just stated concerning the respective curability of the two forms of the disease.

Having said this, a serious error in the interpretation of the idea must be avoided, which would have the effect of exaggerating beyond all truth, or even semblance of truth, the possibility of cure in pneumonic phthisis. Pneumonic tuberculosis (whether tuberculous pneumonia or broncho-pneumonia) is not the only disease which, with an acute outbreak, produces permanent lesions in the parenchyma of the lungs; the simple form of broncho-pneumonia must not be forgotten, which, both in its subacute and chronic form, is characterized by alterations of which the stethoscopic indications are closely analogous to those of tuberculous pneumonia. These conditions may in the end produce true consumption, and may be called on that account phthisigenic (*phthisiogènes* *). It is, however, equally certain that they are in no way connected with the tuberculous diathesis, or with phthisis as the term is understood

* See note, chap. i. p. 13.

at the present time; that they are less serious than tuberculous pneumonia; and should this distinction be overlooked, favourable results may be interpreted to the advantage of pneumonic tuberculosis, which are really connected with simple broncho-pneumonia in its subacute or chronic form.

When the pathology of the disease was considered, the special necessity of making such a diagnosis was mentioned at some length, and this subject will not be again discussed. I need merely state that it is surrounded by numerous difficulties, since no certain or sufficient light is thrown upon it, either by the progress of the disease when it commences, this being acute in both cases, or by the stethoscopic indications, or even by the pathological conditions which may have preceded in either case the pulmonary affection. Measles, in fact, or whooping-cough, may produce either simple or tuberculous broncho-pneumonia, though when the preceding disease is typhoid fever or diphtheria the difficulty is certainly less, both these affections being usually the origin of simple broncho-pneumonia. In many cases the diagnosis may be facilitated by considering the seat of the lesions, tuberculous pneumonia being specially liable to affect the upper lobes, which the simple form is not. This differential character, however, is not absolute, so that in fact the true elements of diagnosis are alone furnished by the family history, the personal antecedents of the patients (as regards scrofula), the presence or absence of indications that the innate or acquired diathesis exists, by the character of the primary focus of disease—since if the whole lobe be affected, the complaint is in all probability pneumonic tuberculosis—by the fact that frequent and extensive attacks of hæmoptysis, which are usually absent in the simple, may occur in the tuberculous form, and lastly by the time within which the state of consumption occurs, this taking place far the most rapidly in the tubercular form of the disease.

Such, in general terms, are the multiple diagnostic indications as regards tuberculous pneumonia; these indications have, I believe, been scrupulously and exactly followed in the cases which I have mentioned as examples of recovery; in these the whole lobe was affected, which, as I have said, is one characteristic of the disease; and, having made the above explanation, I have now no retraction to make from the statements already made as to the curability of this form of phthisis.

Having concluded this necessary digression, which it is hoped will prevent all confusion, the conditions which influence the curability of pulmonary phthisis will be again discussed, and the principal symptoms of the disease will be considered from that point of view. This will be done at the commencement of the next chapter.

CHAPTER III.

THE CONDITIONS WHICH INFLUENCE THE CURABILITY OF PULMONARY PHTHISIS (*Continued*).

Data furnished by symptomatic phenomena—Gastro-intestinal symptoms—Severe forms of laryngitis—Hæmoptysis, with or without pyrexia—Hæmoptysis in pneumonic tuberculosis.

Emaciation—Pyrexia—Importance of the condition known by the name of erethism (*éréthisme* *)—Intercurrent congestion or inflammation.

Preponderating importance of the condition of general health — Individual character of the prognosis—Special effect on the prognosis resulting from the possibility of treatment.

Conclusion—Division of subject.

In continuing to study the conditions which influence the curability of pulmonary phthisis, the symptomatic phenomena of the disease must still be considered. This part of the subject deserves much attention, since in every condition and form of the complaint there are certain symptoms which may diminish or even destroy the least hope of recovery.

In the first place, the serious nature of gastro-intestinal troubles should be mentioned, which results from their early occurrence and persistent character. These disorders increase the organic waste, which interferes with healthy nutrition, a necessary part of the treatment, and prevents the use of certain most useful therapeutic agents, such as arsenic, cod-liver oil, or creasote. So long as they continue, time is use-

---

* The word *éréthisme*, as M. Jaccoud explains in this chapter, means nervo-vascular excitability, giving rise to palpitation or other cardiac disturbance on the one hand, to insomnia and other forms of nervous derangement on the other.

less so far as the treatment of the disease is concerned, and is even harmful on account of the loss with which their occurrence may be associated. Should their duration be prolonged, the results, acquired with difficulty by a treatment of many months, may be lost, the patient being after their cessation in a worse condition than before such treatment was begun. Should the troubles of the digestion become permanently established, should diarrhœa exist, shown by the clinical symptoms to be due to intestinal ulceration, whatever the condition of the pulmonary lesions may be, all hope vanishes, and a decree of absolute incurability must be passed.

The condition is no less unfortunately modified by the development of laryngeal affections, that is to say, when these are severe and persistent. Observation has shown that the chronic laryngitis which accompanies tuberculosis, whatever its nature may be, aggravates the state of consumption, should it already exist, or hastens its occurrence. These symptoms are so much the more formidable, as their development takes place earlier in the disease, since they themselves may precipitate the deterioration of the general health, and remove from the patient those relatively favourable prospects with which the initial period of the pulmonary lesions is usually associated. When occurring thus at an early date, as not uncommonly happens, ulcerative laryngitis is a still more serious complication; and, though having an indirect influence upon the disease, it is quite as fatal in its result, removing from the treatment its most powerful arms at this period, namely, hydrotherapeutics, pneumatic treatment, and residence at a high altitude. In such conditions recovery must be looked upon as unlikely, or it may be impossible.

As regards the prognosis, hæmoptysis has not the constant and invariable signification which is too often attributed to it; without considering that abundant and appalling hæmorrhage, which, at any period of the complaint, and specially where caverns form, may suddenly give rise to a fatal termination, hæmoptysis of moderate severity, occurring at the commencement or during the course of the disease, before ulceration has taken place, should not be looked upon as a sufficient reason for abandoning all hope of recovery. During the initial period, when the danger is specially due to attacks of congestion which cause the lesions to increase in size, the hæmoptysis may be, as it were, critical in character, and dissipate, for a time at least, threatening inflammation. In such cases the event, far from being a sign of incurability, is, on the contrary, a healthful process. Hæmoptysis occurring in other conditions must always cause anxiety, but the opinion to be formed in consequence, with regard to the final issue of the disease, depends upon numerous circumstances. The frequency with which it takes place, its effect upon the pre-existing lesions, which can be ascertained by an examination of the chest before and after its occurrence, the relation which the number and abundance of the hæmorrhages bear to the condition of the organism in which they occur,—such are the most important circumstances, which vary in each individual. Equal anxiety should be felt as to the existence of pyrexia. Hæmoptysis occurring in a patient who is not already feverish may occur without pyrexia, being consequently less serious than in the reverse case. Though other circumstances are the same, it may again be attended by a febrile condition of greater or less intensity, which ends when the hæmoptysis is over, so that the patient returns to the same apyretic condition as before the event took place. This form of hæmoptysis is more serious than the preceding, specially if the hæmorrhage recurs with the same characters; but does not indicate a definite aggravation of the disease, or that the complaint is incurable.

In support of this assertion, the remarkable case of Prince Ou—— may be quoted. This patient, whose age was forty, with tubercles which were softening in one of the upper lobes of the lung, was affected almost immediately after his journey from Méran to Paris by severe hæmoptysis. On this and the following day his condition passed from an apyretic state to one of fever, the temperature at night being always above 102° F. (39° C.), and frequently reaching 104° F. (40° C.). He remained in a serious condition for fifteen days, after which the hæmorrhage ceased, the fever entirely subsided, and, after a relatively short period of convalescence, his recovery was so far complete that he could undertake the return journey to Méran without more danger than he had incurred when visiting Paris. One fact should be noted, which is not the least interesting in the account of this case, namely, that this was by no means the first time that hæmoptysis had occurred. In the four previous years he had had four attacks, usually in the spring, and always when some change of residence had caused strong to be substituted for weak atmospheric pressure.

According to the report of the physicians to whose care he was then entrusted, these attacks were neither less severe, nor attended with less pyrexia than when Paris was visited. Notwithstanding their serious character, however, they had no appreciable effect upon the evolution of the disease. Whilst the hæmoptysis continued, the patient was certainly in imminent danger, but when the attack was over, the affection returned to the inert condition in which it had now been for many years, and which constituted a *statu quo* equivalent to relative recovery.

The curability of the disease is much more seriously compromised if the pyrexia aroused by hæmoptysis continues when the latter has ceased, and becomes persistent. Of whatever type the fever may be, it modifies seriously the condition of the patient, for, as will soon be explained, whilst it lasts every idea of improvement must be abandoned.

Lastly, it should be pointed out that the hæmoptysis

which occurs at the commencement of tuberculous pneumonia may be most severe. When small in quantity and of short duration, it gives no indication as to the subsequent progress of the disease, but at times the initial hæmoptysis of tuberculous pneumonia continues unchecked for many days, and in such abundance as to cause anxiety from that fact alone. The hæmorrhagic form of the disease then exists of which the prognosis is most unfavourable, not only from the fact that sudden death may occur, but because in case of survival there is but little hope that cure or even improvement will take place in the pulmonary lesions. At the same time, notwithstanding the initial hæmorrhage, a favourable evolution of the disease is not impossible, and the account of such a case has been already given, occurring in a man aged forty-two years, who recovered at the ulcerative period of the disease. Such facts, however, are very exceptional.

Emaciation is a symptom of equal importance with respect to the question of recovery. If it is but temporary, and its occurrence can be plausibly explained by the existence, either of an acute complication, continuous pyrexia, or some temporary digestive disorder, it has not an absolutely unfavourable signification, and merely implies delay in the process of recovery, supposing this to be possible. Should emaciation, however, occur at an early period in the disease, persist, and perhaps increase, whilst the complaint, judged by the character of the local lesions and the absence of fever, seems to be stationary, it is then truly consumptive in character, and excludes the possibility of cure. In every condition of the disease the first indication of temporary or permanent improvement is decrease in the emaciation, and increase in the weight of the patient.

The character of the prognosis is no less influenced by the condition of the patient as regards pyrexia. In chronic forms it is the early appearance, in pneumonic forms the long duration of this symptom, which is of such importance. As long as fever exists, no improvement can take place, so that the physician must consider it his first duty to combat this symptom, notwithstanding its purely symptomatic character. Criticisms have been often passed upon the denominations active or florid, as opposed to passive or torpid phthisis. These criticisms are unjust; for if by the term florid is meant, as it should be, phthisis attended usually with fever in an excitable person, and by torpid phthisis a form of the complaint which is generally apyretic in a non-excitable patient, there is no distinction in the clinical history of the disease which is more legitimate and, at the same time, of greater service as regards both the prognosis and treatment. The terms, however, should be employed with this precise signification.

There is thus another symptom, due to the individual character of the patient, which, though less important than pyrexia, should be taken into serious consideration in determining the question of curability, namely, the nervo-vascular excitability, to which the name of erethism (*éréthisme*) has been applied. This condition may produce mere cardiac excitability, and frequent attacks of palpitation in consequence. In other cases there is general nervous excitability without any pronounced visceral disturbance. In whichever form, however, the condition occurs, the result is the same, and the prospect of recovery is diminished by its presence for reasons of different kinds. Symptoms which are themselves serious, such specially as pyrexia and hæmoptysis, may be produced or continued owing to its existence; at other times, by producing waste, it adds to the constitutional loss which characterizes the disease; it may, again, prevent the adoption of certain therapeutic means of treatment, such, for instance, as residence at a high altitude, or the use of hydro-carbon preparations as internal remedies. So long as the condition of erethism persists, such difficulties will continually arise, and but too often it is necessary to disregard the fundamental indications of the disease in order to counteract, and remove if possible, this persistent excitability, which should be looked upon as a true complication of the complaint. Though the general condition is satisfactory, though the lesions are circumscribed and stationary, the condition of the patient may be so serious, owing to this fact, that the idea of improvement must be definitely abandoned.

With regard to the prognosis, and therapeutic difficulties connected with the disease, this special condition should be better known. Amongst other cases which support these assertions, that of an American lady, sent by Professor Wyss of Zurich to Paris, may be mentioned. In this patient the tubercular lesions, although bilateral, were circumscribed in size; there was no emaciation, there were no laryngeal or intestinal complications, or daily pyrexia, and, in short, the disease presented in character the most reassuring simplicity. Different, however, indeed did this appreciation of the condition become when the person was considered rather than the disease. Exhausted by labours, and suckling repeated at short intervals, worried by family cares, this patient was from the commencement of the pulmonary affection in an extremely excitable state. Continuous insomnia, frequent attacks of palpitation, constant acceleration of the pulse without rise of temperature, and such gastric irritability as at times to prevent all administration of food, were the consequence of this complication. The medical treatment which appeared most suitable failed to modify this state, and the most appropriate remedies would at times aggravate the condition; milk, which was first taken, diminished to a notable degree the irritability of the stomach, but the other symptoms continued without change, counteracting the effects of whatever treatment might be employed to arrest the disease. Against medical advice, the winter was spent at one of the medical stations in the

Riviera. The nervous symptoms became more serious; no remedies, and scarcely any food, could be taken except milk, and emaciation rapidly ensued. No power of resistance now remained, fever ensued, and in March the patient was in all respects in a more serious condition than when she had left Paris. The state of nervo-vascular excitability (*éréthisme*) persists, and consumption will undoubtedly close the scene.

This, then, was a patient in whom the tuberculosis considered alone was not specially formidable, whereas the nervo-vascular excitability produced the most harmful effects, such as fever, constitutional waste, and inability to respond to treatment; to this unfortunate condition absolute aggravation of the disease was due.

The condition produced by nervous erethism (*éréthisme nerveux*) is not always of such a serious character, but in every case furnishes an important element of consideration with regard to the curability and treatment of the disease, and should never be disregarded.

The variable effect which the production of caverns has upon the process of recovery has been already considered, and this need only be mentioned for the purpose of recalling that their existence is an important consideration with respect to the question now under consideration.

As far as the causes, lesions, and symptoms of the disease are considered, these are conditions which influence either favourably or unfavourably the curability of pulmonary phthisis. At the same time, though such data are important, they occupy but the second place in the delicate appreciation of the probability or possibility of recovery, whether absolute or relative. The effect of each and all depends upon two other considerations, which are the fundamental elements of that vital problem which every patient affected with phthisis presents to his medical attendant. These two important considerations, as has already been indicated, are firstly tho general condition of the patient, and secondly the existence of or freedom from intercurrent attacks of congestion or inflammation.

Whether tubercular in nature or not, each of such acute complaints is certain to aggravate the disease, and, whether congestive or inflammatory, to hasten the evolution of existing lesions. Even congestion gives rise to the development of similar lesions, and in any case, should neither of these consequences ensue, the acute complication by which the patient is affected causes the process of organic consumption to be increased. Though other conditions presented by the patient may be favourable, as far as recovery from the disease is concerned, these attacks, if frequently renewed, are themselves sufficient to overthrow any chance of a favourable termination.

So far as the general state of the patient is concerned, the so-called constitutional condition, this is of more importance than any of the symptoms mentioned. The ultimate evolution of the disease, and the whole prognosis as regards curability, depend upon the degree of existing malnutrition when the disease first appeared. Nor could it be otherwise, for upon what else could the whole question really depend? Since the organism is affected by a disease which is eminently consumptive in character, and receives a shock both severe and of long duration, it is most important that it should have power to resist its effects; nor does this suffice, since repair has to be effected, which can only be done by the vital energy of the patient. Thus it is the strength of the organism which both resists and repairs the effects of the disease, and in this way the degree of malnutrition furnishes in each patient an exact measure of his condition and of the favourable or unfavourable prospect of the disease. Though other circumstances may be most favourable, the state of the case is not altered, and if once the constitutional condition, the nutrition, is affected beyond all hope of recovery, the issue of the disease is certain, and all hope of cure must be abandoned.

Pathological anatomy leads to the same conclusion. The harmful and destructive process is that of caseation, which is essentially the passive result of debility. The process of repair, on the other hand, is the fibroplastic process, due to irritation of healthful nature, which requires and demands, as does all work of sclerous formation, pronounced activity both in the local and general nutrition. These seem to be facts which cannot be denied. Whether, therefore, the results of clinical observation be considered, or the anatomical changes which accompany the favourable or unfavourable evolution of the disease, the same conclusion is reached with respect to curability, namely, that the condition of organic strength and the degree of malnutrition overbalance all other elements of judgment.

The above considerations also show that insufficient and improper nutrition are the origin of tubercles and of their evolution, and that such malnutrition is the first source whence both the prognosis and the therapeutic treatment should be deduced. All other considerations are of secondary importance.

An interesting sequence of this fact should be mentioned. The constitutional state is an individual condition, and the conclusions which may be drawn from its character vary according to the patient. They depend upon how the general condition is as regards the nature, previous progress, and probable evolution of the pulmonary lesions. Hence it results that the fundamental element of prognosis is derived from the patient himself, and not from the general indications furnished by the cause of the disease, its symptoms, or any of the above-mentioned conditions. It is true that, considered in a theoretical and abstract sense, acquired phthisis is less severe, and more curable than the hereditary form. Notwithstanding this general fact, however, it may happen that a patient suffering from acquired phthisis is more seriously affected and in a less curable condition than one suffering from the hereditary complaint. The so-called arthritic form of phthisis, again, considered in the same general way, is the least serious of all the secondary forms of phthisis, but in a patient with this disease the sentence of incurability may be pronounced more early than in one who is suffering from some other form, which is usually and with reason considered more severe. Hæmoptysis, intercurrent congestion or inflammation in every form of the disease, modify the actual and subsequent condition of the patient. At the same time, two patients who have suffered in a similar way from these formidable complications may be, as far as the subsequent evolution of the disease is concerned, in a very different state owing to their respective constitutional conditions. The Russian prince, whose case has been already mentioned,* withstood the effect of repeated hæmoptysis attended by pyrexia, which would have been fatal to many patients, nor did the progress of the disease seem to be hastened by its occurrence. The young man from the same country, who has already for some years enjoyed the benefit of relative cure, was enabled by his constitutional strength to withstand repeated attacks of broncho-pneumonia, which might have been fatal to a weaker patient.

It is needless to multiply examples of this fact. The same is true with regard to each element of prognosis in this disease. At all times and in all conditions of the illness the constitutional strength of the patient holds the first place in importance, and without denying in any way the value of the general sources of judgment, the appreciation of the curability of the disease should be looked upon as a purely individual problem, which repeatedly arises with respect to each patient. The prognosis as derived from the pathology of the disease is of but secondary value.

One other condition affects the curability of the disease,

* See p. 49.

namely, the treatment. Not only should the therapeutic means which are adopted be appropriate, and perseveringly carried out, but the possibility of their employment must also be considered. Nothing is so complex, so difficult to conciliate with the ordinary habits of life, so costly in expense, as the treatment of pulmonary phthisis. But too often considerations connected with the family or the fortune of the patient, to which the physician must yield, stand in opposition to his wishes, and too often in consequence a palliative form of treatment must be adopted in the place of one which might lead to recovery. The disease may thus remain incurable, not on account of its own serious nature, but because it cannot be completely and constantly resisted. This difficulty is always encountered, and the disease is therefore quite different, as it affects the rich or poor, so far as the result of treatment is concerned. The disease in the latter case is not itself more serious than in the former, but those who are affected by it have not the same power to employ all the means by which it can be resisted. Thus, if patients of this class are alone seen, the logical deduction but false conclusion is likely to be made that phthisis is incurable, while a protest is made against the error of those who form a more favourable opinion. There is no mistake on either side, but the field of observation not having been the same, a difference of opinion exists which does not disprove the notion that the disease is curable. This mistake, however, is frequently made, and implies thoughtlessness, to use no stronger term, on the part of the observer. When it is asserted that pulmonary phthisis is curable, this proposition applies only to cases in which the disease receives appropriate treatment, and the numerous cases which are incurable, from the impossibility of such treatment being adopted, cannot in any way diminish the truth of this affirmation.

Though in private practice a great difference exists as regards the curability of phthisis, this fact is much more evident in hospital patients. In them the conditions are most unfavourable; the very idea of recovery must be looked upon as a chimerical dream, observation showing not only that the disease ends fatally in all whom it strikes, but that its duration is shorter than in other classes. Before visiting the hospital, these patients are more seriously affected than others on account of the deplorable hygienic conditions of their life, and the treatment adopted after their admission can only be palliative. Being in a hospital ward, no curative treatment of any kind can be adopted, for without mentioning the employment of mineral waters or climate, which is of course impracticable, the hospital life of consumptive patients in large towns is incompatible with the fundamental means of treatment, namely, pure air, the sun, and a country life. It is not therefore surprising that phthisis in these cases should be nearly always incurable, since both the state of the patient and the conditions in which he is placed must cause the disease to be more unfavourable in its effects, and to terminate with greater rapidity and a more fatal result. Yet in this case again it would be a mistake to generalize, since the conclusion is only true as regards a certain class of patients. These facts, however, prove that the mode of treatment adopted should be looked upon as the most important condition which influences the curability of pulmonary phthisis.

This preliminary question, usually but little considered, is of such importance as amply to deserve the extended consideration which it has received. It is evident that the possibility of treatment must be the necessary and indispensable introduction to its form, since a belief in the possibility of cure can alone engender and sustain the persevering course of action required.

The treatment of the disease, properly so called, will now be considered; that is to say, the means which may be employed with the object and effect of conferring on the patient the benefit of that absolute or at least relative cure which has already been mentioned.

This part of the subject will be considered as follows. In describing the pathology of the disease it was divided into: (1) Chronic tuberculosis or ordinary phthisis; (2) pneumonic tuberculosis or phthisis; and (3) acute miliary tuberculosis or granulosis. Whether the disease be regarded clinically or anatomically, this fundamental division must be made, and in considering the therapeutic means of treatment, the same plan may be legitimately adopted, and will serve as a guide in what is still to be discussed.

Before, however, entering into the treatment of these different forms, the prophylactic treatment of pulmonary phthisis must be considered, and with the full details which its importance requires. The object of such treatment is to counteract the phymatogenous* tendency or diathesis, and thus to prevent any manifestations of the tubercular affection; it is adopted during the premonitory period of the disease, whatever form the complaint may be likely to take should it become developed. This is, therefore, one among many reasons for the importance of such treatment; tuberculosis in general being treated, and not one merely of the clinical forms in which it may show itself.

This having been done, the ordinary form of phthisis will be considered, characterized as it is anatomically by the gradual evolution of tubercles, and the treatment of the confirmed disease will be dwelt upon at its commencement, and subsequently during its stationary or active periods. The discussion of pneumonic phthisis will follow, specially of the acute phase which characterizes its invasion, since the treatment of the disease subsequently, should the patient reach such a stage, is the same, except as regards the climatic stations, as that of the ordinary disease. Lastly, the treat-

* See note, chap. i. p. 7.

ment of acute miliary granulosis will be mentioned, a form of the disease to be rapidly considered, firstly, because it differs clinically from the forms of tuberculosis commonly designated by the general term pulmonary phthisis, and secondly, because a few propositions will suffice to make known the principles and mode of treatment to be adopted in the presence of this disease.

In the course of this discussion the hygienic and medicinal treatment will alone be considered, the indications for the employment of mineral waters or climatic stations being subsequently discussed.

CHAPTER IV.

PROPHYLACTIC TREATMENT.

Prophylactic treatment—The indications which it follows—Its value—Conditions of success.

Signs which indicate the utility of prophylactic treatment—Indicative signs furnished by the ætiology of the disease—Hereditary transmission—Constitutional debility—Caseous remains of scrofula—Self-infection—Measles, whooping-cough, pleurisy—Hæmoptysis; phthisis produced by hæmoptysis—Transmission of phthisis: experimental transmission; spontaneous transmission; contagion—Modes of transmission: inoculation inhalation, alimentation—Practical consequences as regards the surroundings of the patient.

Indicative signs furnished by the patient—External appearance—Anæmia, obstinate dyspepsia, etc.—Position of the clavicle—Diminution of respiratory force—Conclusion.

In conformity with the plan already mentioned, the prophylactic treatment of pulmonary phthisis will be discussed in this chapter. It is unnecessary to dwell at any length upon its preponderating importance, and it will be sufficient to consider in this respect the special condition of the patients affected by the disease. As yet there is no local lesion, no pulmonary trouble has excited suspicion, but on account of unfavourable family antecedents, or from innate or acquired debility, a condition of constitutional malnutrition has been developed well adapted to the existence of phthisis, and causing its development to be feared at some later date. Persons in this condition are not ill, but predisposed to disease. It is not actual treatment which they require, but prophylactic care; and medical intervention is the more efficient, since it is not local changes which require attention, these having as yet no existence, but the constitutional condition to which these lesions are due. Since there is but one indication of treatment, it is the more easy to follow; while it may also be said that this indication is connected with the cause of the disease; the most important and fundamental consideration. Prophylactic treatment can, therefore, attain its object with relative facility from the fact that it has but one aim and complete freedom of action, while, should its end be gained, the disease would be crushed before its menaces have become an accomplished fact. It is, therefore, certain that this part of the treatment is of greater value, and has more power than that which is subsequently adopted.

As already said, it has a general effect, being applied to pulmonary tuberculosis in its entirety, and not to one or other only of its clinical forms, a circumstance which establishes the incontestable superiority of such preventive treatment.

Another fact, however, is no less evident, namely, that on account of its general object prophylactic treatment is powerless without the help of time. Restoration, transformation of the constitution, cannot be accomplished in a few days. It is a work of long duration, and the efficacy of prophylactic treatment, which is theoretically absolute, really depends entirely upon its being continued for a length of time, or, in other words, it is the more powerful in proportion as it is begun more early in life. Hence arises the practical importance of the following question, which should be solved from the very first, namely, By what means can the physician know in any case that such prophylactic treatment of phthisis should be adopted? Everything depends upon the answer to this question, and upon the solution of this problem at a sufficiently early date, the success or failure of treatment during the premonitory period, but too often neglected, entirely depends.

That the question may be answered satisfactorily, it will

not suffice to take into consideration the individual conditions of the patient, whether general or local. Besides this, and before all, perhaps, great attention should be paid to the ætiological data of the disease; and in many cases, whatever the condition of the affected person may be in other respects, the existence of these data will show the necessity of prophylactic treatment, and in many cases will indicate how it should be employed.

In the first place, then, those ætiological conditions will be considered which present such an indication.

The existence of phthisis as a family complaint should first be mentioned, that is to say, hereditary transmission either direct or collateral, the signification of these two conditions being by no means identical. Strictly speaking, if the indications of collateral transmission are furnished by a distant branch of the family, or again, if the individual concerned is undoubtedly possessed of a vigorous constitution, premonitory treatment may not be necessary. Direct hereditary transmission, however, or rather the fear of such transmission, is of itself and in all circumstances enough to render its employment necessary, whatever may be the constitutional condition of the person, the most active treatment being necessary should other children have been already affected by the disease. The chance of such direct hereditary transmission is the most unfavourable condition as regards prophylactic treatment, and too often the best directed and most persevering efforts are fruitless, or the onset of the disease is simply retarded by them, if such a statement can ever be made with certainty in this disease. It is a fact which, having been proved to be true by the experience of centuries, obliges the physician to forbid marriage in the case of those who, on account of such direct inheritance, are threatened by the disease, the danger of its appearance being always present. The prohibition seems, perhaps, cruel, but as the result of enlightened and approved clinical experience should be rigidly imposed, tuberculosis not diminishing in gravity as it passes from one generation to another.

Though having by no means the same constant and serious consequence, other circumstances connected with the family may be looked upon as indicating the advantage of premonitory treatment. The old age of the parents, whether at or before the proper time of life, the existence in one of them, or even more so, in both, of alcoholism, scrofula, or diabetes mellitus, and lastly a line of consanguineous marriages carried back from parents to ancestors, are the most important of such conditions.

It is needless to dwell upon that form of constitutional debility which is innate or natural to the individual, or upon the weakness which results from enforced study in early life, from physical exertions ill adapted to the age of the person, from confinement in a crowded atmosphere, a sedentary life in dark and ill-ventilated dwellings, insufficient nutrition, or prolonged lactation. It has already been mentioned, in the general observations which were made as to the ætiology of the disease, that constitutional malnutrition, whatever may be its origin, is a cause of the complaint. It is therefore evident that this should be opposed, affording as it does a good opportunity for the adoption of prophylactic treatment.

There are other considerations, however, connected with the ætiology of the disease which should be discussed at greater length.

Scrofula in infancy and youth is one of the most important of these, and should cause a watchful care to be kept by the medical attendant over those who have been affected by it. Connected also with debility and with constitutional malnutrition, it presents an undoubted connection with phthisis, a fact which is the more evident since the existence of the massive pneumonic tubercle has been recognized. In such conditions other questions should be considered. If scrofula is still an active disease when the age is reached at which phthisis is specially liable to occur, the necessity of prophylactic treatment is so imperious that it may be looked upon as part of the curative treatment which should be adopted. It may, in fact, be regarded as almost certain that in such a case scrofula will end in the formation of tubercles, and that the disease long considered possible, not to say imminent, will be brought into existence.

In other cases the manifestations of scrofula may have ceased when the age is reached at which the onset of phthisis is to be feared. The prospects are then more favourable; but before the patient can be said to be in perfect safety, the way in which the cure was effected and the degree which it reached should both be taken into consideration. The recovery may be complete and no indications of the disease remain, in which case, should the constitutional state of the person be in other respects satisfactory, the condition is good, and imposes no necessity for prophylactic treatment. Often, however, the recovery is relative, and not absolute; the active effects of scrofulous disease are arrested, and this affection may be looked upon as cured. There remain, however, more or less numerous changes due to its previous existence, such as glandular or other tumours, now in an indolent and inactive condition. The condition of things is then quite different from that which exists in the previous case, and far more unfavourable.

Dreaded as it would be on account of the previous existence of scrofula, tuberculosis is rendered even more imminent by the special danger which attends these foci, when inflammation has been replaced by caseous degeneration. This danger I have designated by the term "tubercular auto-infection" (*auto-infection tuberculeuse*). Such secondary infection, which to a certain extent justifies the old idea that scrofula was the cause of tubercle, results from the more or less active absorption which takes place in these remaining scrofulous deposits. The existence of such infection, which has been known to occur since the time that the works of Dittrich, Virchow, Buhl, Lebert, and Wyss were written, has been often experimentally proved by the inoculation of caseous products, whether pulmonary, pleural, or glandular. In some cases tuberculosis is produced in the lung alone, in others throughout the body, the lymphatic vessels and veins being the channels by which the infection occurs. By one of these channels the infecting particles reach the large venous trunks and the right side of the heart, and are thence discharged into the pulmonary artery and lungs by the mechanism of capillary embolism. This diffusion remains confined to the lungs so long as the capillary vessels, which are intermediate between the pulmonary artery and veins, oppose a barrier to the farther passage of these infective particles; should they not do so, and this obstacle be passed, the pulmonary veins and left side of the heart are reached, and the particles may be disseminated throughout the organs and tissues by means of the arterial branches. This fact is borne out in the works quoted above, as well as by the more recent observations of Orth and Huguenin; and the researches of Ponfick seem to show that the thoracic canal is itself altered in cases of generalized miliary tuberculosis produced by artificial means. Whenever the granulosis is localized the walls of this canal remain apparently unaffected, but in the diffused form a confluent eruption of small tubercular nodosities is almost invariably found in the interior of the vessel springing from its wall, and attributed by Ponfick to the passage of lymph which contains a specific irritant.

The mode in which the infective matter coming from a local focus of disease is diffused throughout the organism is certainly of a most complicated nature, and in the present case may be termed hypothetical, since the supposed embolus, by means of which the disease is spread, cannot be perceived more than in other cases of capillary embolism. This objection, however, to the theory of its existence has no foundation; firstly, because secondary tubercular infection due to the existence of caseous foci is a fact which has been definitely established, taking place necessarily as above mentioned; secondly, because the same has been shown to be true in the case of larger particles, the long and tortuous passage of which is far more surprising on account of their larger size. The case reported by Soyka, in 1878, which I should not leave unnoticed, is most convincing with regard to this point. In a man aged seventy this observer found an infiltration of carbon particles at the apex of both lungs; the lymphatic vessels of the pleura and lung were tinged of a bluish-black colour by the accumulation of these elements, which were also found to be abundant in the bronchial glands. There was nothing specially surprising in this fact, but particles, identical in nature, existed in the liver, spleen, and kidneys. In the first of these organs they were found in connection with the bloodvessels, either in the tunica adventitia, or surrounding connective tissue. The hepatic cells did not seem to be invaded, but the particles existed between them, and were found in greatest number at the margin of the columns. This fact affords conclusive proof of the migration of infective particles, and that they are not developed in many places at the same time. When particles of carbon are found in the liver, spleen, and kidneys, this is due to their passage from the network of pulmonary arteries into which they were brought through the lymphatic vessels and right side of the heart, to the corresponding pulmonary veins, the left side of the heart, and thence into the different arteries of the body. This is the same course as that followed in tuberculosis, being perceptible in the above case on account of the nature of the displaced particles.

The difficulty in making such a long circuitous course cannot, therefore, be looked upon as a reason for disbelieving the secondary infection produced by caseous foci. This, as above stated, has been experimentally proved to occur. The pathological facts which establish it almost conclusively prove its existence. In a patient who has died from the effects of miliary granulosis of the lungs, either pulmonary or diffused inflammatory foci of ancient date and at the period of caseation are found in the lymphatic glands or serous membranes. The two lesions cannot be regarded as cotemporary, the caseous foci being certainly and by far the most ancient; and since it has been experimentally proved that the inoculation of such products may be followed by a general formation of granules, it must be admitted that in this case absorption has taken the place of inoculation, and that the patient has infected himself by means of the diffusion of harmful elements.

Such facts are not of rare occurrence; Buhl, Dittrich, Lebert, have mentioned them again and again, and I myself have seen many clear examples of such infection. Tuckwell, in the autopsy of eight cases of tubercular meningitis and of generalized miliary granulosis, found in each patient ancient scrofulous foci as the primitive lesion, these facts leading him to the conclusion that all such foci are the generative causes of miliary tuberculosis, encysted foci being alone excepted. Mazzotti, in ten cases of miliary granulosis, found that in four caseous foci were the origin of the complaint, in three a focus of suppuration. Duckworth and Gee have reported isolated observations of the same kind, and it would be useless to say more of this pathogenic connection, which cannot at the present time be seriously called in question.

The fact, then, may be looked upon as certain, that the caseous remains of cured scrofula are an efficient cause of subsequent tuberculosis, and should therefore be regarded as positively indicating the necessity of prophylactic treatment.

More than this, such conditions point out the special direction which such treatment should take, the prevailing indication being to remove these foci which at any time, even after a long period of inertness, may produce tuberculosis.

The interest of these facts of auto-infection, which is considerable, is not limited to scrofula and its caseous foci, but extends also, as will be seen, to certain forms of pleurisy. In these cases the pleurisy is usually of long duration, the effusion being moderate in amount, and the disease ending in the formation of adhesions. Though the effusion is removed, however, the patient does not recover; he remains indisposed, and after a time, varying from a few weeks to a few months, at the spot where the adhesions formed a pneumonic focus is found to exist, which is soon shown by its evolution to be caseous in character. This focus exists alone, or other foci may form at a greater or less distance from it; and should the patient succumb, all the particulars described above are verified, namely, the absence of effused fluid, the presence of adhesions, and the existence of pneumonic infiltration, either consolidated or already softening in that part of the lung where the adhesions have formed, and alone or with disseminated foci. I have myself seen some ten cases of this kind; and Knævenagel, the distinguished physician of Cologne, who has also investigated this question, reports numerous examples of its occurrence. The seat of the pleurisy is notable; in most cases the antero-inferior or middle region of the left side of the chest is affected, which, in fact, according to Knævenagel, is constantly the seat of the disease. This, however, I cannot admit, since in two of my cases pleurisy existed posteriorly and in connection with the lower lobe of the right lung, though in other respects the progress and course of the disease were in all respects as mentioned above.

Knævenagel is not convinced that these secondary pneumonic foci are of tubercular nature, which he seems even disposed to deny, in his work entitled "Contributions to the local development of Phthisical Conditions in the Lung." For my part, however, I feel able to affirm the tubercular nature of these forms of pneumonia. Already three times—and this recalls me to the real subject of our consideration—already three times I have been able to recognize, independently of foci which were relatively of more ancient date, the formation of new granulations, which I thought it right to attribute, for reasons previously given, to the mechanism of auto-infection (*auto-infection*). The identity of their nature seems to me, therefore, certain. I would mention the most recent of these cases with some detail. It relates to a man aged twenty-six who came under my care on the 29th of October, 1879, and whose autopsy was performed in January, 1880.

As does not usually happen, pleurisy with adhesions existed on both sides; the left anterior region was affected in the usual way, but the adhesions were most marked on the right side, where what might be termed a symphysis had formed, uniting the diaphragm to the ribs around the whole circumference of the chest. On the left side such a symphysis only existed anteriorly, being formed by membranes of such thickness that during life there was no trace of normal resonance in the so-called semilunar space (*dans l'espace dit semilunaire*). Each lung presented internally at its base a compact focus of infiltrating disease, consisting of homogeneous caseating substance, resembling in all points a mass of yellow Roquefort cheese. On each side of the lower lobe were seen anteriorly, and immediately beneath the pleura, two or three foci of disease, extremely small, isolated from each other, and resembling those mentioned above, with which they were evidently cotemporary. So far there was perfect similarity between the condition of the two lungs, but around the large focus of disease on the left side was found a zone of grey granulations with a marked yellow hue. The number of these granulations diminished from above downwards, and from the centre to the periphery in proportion as the distance from the focus of infiltration became greater, but even at the apex some granulations could still be seen in places. In the right lung none of these lesions could be perceived. This case seems to prove the possibility of secondary granular infection by diffusion from a caseous focus of pleural origin. The possibility of such a connection is in fact so well established, that these forms of pleurisy should be placed among the conditions which may end in tuberculosis, and which therefore indicate the employment of prophylactic treatment.

Measles, whooping-cough, typhoid fever, and diphtheria, from the effects of which the lung has not entirely recovered, specially the two latter diseases, are more often followed by broncho-pneumonia in its subacute or chronic form. When, however, the hereditary predisposition to phthisis exists, whether innate or acquired, these diseases may be the origin of pneumonic tuberculosis. The pulmonary lesions which follow these complaints should, therefore, be regarded with suspicion and reasonable anxiety, and impose the necessity of active intervention.

Hæmoptysis is another occurrence which may reveal the danger of pulmonary phthisis, that is when it takes place independently of any existing or appreciable pulmonary change. Such hæmoptysis may, in my opinion, produce pulmonary phthisis, and may not only be the effect, as in the large majority of cases, but also the cause of the disease. This fact is certainly exceptional; but rareness does not mean impossibility, and though this idea has been severely criticized, I still believe that phthisis ab hæmoptyse (*phthisie ab hæmoptoë*) has a real existence; and the denial of this fact, deduced as it is from a theoretical and general point of view, cannot efface from my mind the impression of facts which I have myself observed. No fresh example of such phthisigenic hæmoptysis (*hémoptysie phthisiogène*) has certainly come before me since my "Clinical Lectures delivered at the Lariboisière Hospital" were published. In this there is nothing surprising, as such facts are of rare occurrence; and since that time Green and Sokolowski have reported similar cases, while Hohenhausen has made experiments, which, though they are not conclusive, still prove that blood has a harmful effect upon the alveoli and bronchial tubes, which renders them far more apt to be affected by morbific agents.

Considering the object of our present studies, a very brief account of these experiments will suffice. If air already rendered septic by being passed over wounds or ulcers is introduced into the lungs of healthy animals, no inflammation or other appreciable effect is produced; if, however, before the inhalation of such air healthy blood is injected into the respiratory tubes, such inhalation is invariably followed by pneumonic changes. It may, therefore, be inferred that the presence of blood modifies the surface of the bronchial tubes and alveoli, and renders it apt to be impressed by irritant gases which previously had no morbid effect.

Without exaggerating the importance of these experimental facts, it may obviously be concluded therefrom that blood has an injurious effect upon the lining membrane of the bronchial tubes, and renders it more sensitive to the different causes of irritation. The harmlessness of its presence, which is the most powerful argument against the doctrine of Morton, cannot, therefore, be accepted as an absolute and constant fact.

On the other hand, the observations of Thompson, though not immediately relating to the development of phthisis, show that the effusion of blood into the pulmonary alveoli may leave a permanent impression, and in some cases lead subsequently to the formation of caverns by producing inflammation in the adjoining tissues, and elimination of the nuclei of connective tissue which resulted from the stagnation of blood. With regard to tubercle, properly so called, Thompson does not deny that its formation may possibly follow the occurrence of hæmoptysis, but limits his statements on this point to the assertion that fresh observations are required to determine whether inhaled blood can produce tuberculosis without the intervention of some secondary process. This theory suggests a possibility which is not mentioned above, since in my opinion hæmoptysis can only lead to the existence of phthisis by the intermediate effect of secondary inflammation.

Such is the present condition of this question, and recent observations justify the opinion expressed above, and afford no reason for abandoning the conclusion to which I had been led by my own observation and that of others. I maintain, therefore, that, independently of pre-existing tuberculosis, the occurrence of hæmoptysis should be regarded as one of the circumstances which may lead to the development of phthisis, and which, therefore, indicates the necessity of prophylactic treatment. This necessity is the more imperative when the person affected by hæmoptysis is of the age at which tuberculosis is most apt to develop, and presents the organic conditions which characterize predisposition to that disease.

These presumptive signs, however, do not exist in all cases, hæmoptysis of traumatic origin having been followed on more than one occasion by the development of phthisis in persons apparently possessed of a faultless constitution.

One other ætiological datum must still be considered, which may also furnish useful indications as to prophylactic treatment, and which is in all cases of the greatest interest, namely, the transmission of the disease by infection. The question still to be discussed is whether, in appreciating its curability, the chances of contagion to which a person may be exposed should be taken into consideration.

As long as the experimental transmission of tuberculosis

was only obtained by means of inoculation, this occurrence could be said to supply no knowledge which might be applied either pathologically or clinically. When, however, the subsequent experiments of Villemin and Chauveau had shown that the ingestion of phthisical expectoration might be followed by the same consequences as inoculation of tubercular matter; when the researches of Klebs as to the effects produced by absorption of milk supplied by cows affected with phthisis enabled him to conclude—that milk can produce tuberculosis in the different animals; that the first effect of such tuberculosis is the occurrence of gastro-intestinal catarrh, which is followed by a tubercular affection of the mesenteric glands, and then by tuberculosis of the liver and spleen, and lastly, by diffused miliary tuberculosis of the lungs; that such tubercular infection can be resisted by a powerful organism, and that in such a case the already formed tubercles may heal by cicatrization;—when these facts were shown to be true, it was necessary to reopen the question, and to consider what opinion should be held of the absolute quiescence supposed possible by the classical doctrine of non-transmission of the disease. This question was the more pressing since at that time, namely, between the years 1869 and 1874, observation confirmed by experimental proof enabled Ullersberger, Boineau, Castan, and Weber to furnish convincing examples of such spontaneous transmission.

Although the date is not far removed from our time, the demonstration of this important fact has since then been fixed upon a more extensive and solid basis, both by means of experiment and clinical observation.

With regard to the first, without considering the researches of Biffi and Verga on the one hand, and of Semmer on the other, in which subcutaneous and intravenous injections were made of sputa and other tubercular products, of blood or milk taken from a cow affected by tuberculosis, as having no medical application, the experiments performed by Tappeiner in 1877 and 1878, by a method which rendered the results obtained of undoubted service in pathology, should still be mentioned. With the sputa obtained from phthisical patients and water, he prepared a fluid emulsion, and during an hour and a half on each day dogs were made to inhale this liquid in a pulverized state, these animals being chosen on account of their slight tendency to be affected by tuberculosis. Eleven animals were treated in this manner during a length of time varying from three to six weeks. During the time of the observation the animals preserved a healthy appearance and did not lose weight, but at the autopsy they were all found, with the exception of one doubtful case, to be affected by miliary tuberculosis in both lungs, while in the greater number granular deposits were also found in the kidneys, liver, and spleen. On the evening of the day which preceded that of the autopsy, Tappeiner mixed finely pulverized carmine with the liquid inhaled, and numerous stains of this colour were found at the surface of both lungs. Whilst conducting these experiments by the method of inhalation, other dogs were supplied with food mixed with expectorated products having the same origin; most of the animals escaped infection, but in a small number general miliary tuberculosis was produced, the intestinal tube being the part which specially suffered.

These observations show that the inhalation of tubercular particles has more power to produce the disease than their ingestion. On account of the information furnished thereby, these experiments are of the greatest interest with regard to transmission of the disease from man to man, which ordinarily occurs by way of inhalation. They also explain the difference in the results obtained by those observers who merely tested the transmission of the disease by way of food, which in no way prove that the disease is not transmissible, but merely that the different modes of infection have an unequal degree of power in this respect.

The researches of Schottelius and Corning, which were undertaken in 1878, that is during the same period, should be compared with those of Tappeiner, the results in many points being similar in the two cases.

In continuing this inquiry, clinical facts must also be taken into consideration. In addition to the writings already mentioned, there are the facts stated by Weber and Rohden with reference to the transmission of phthisis from husband to wife, and the cases which induced Webb to give an affirmative answer to the question forming the title of his work, "Is Pulmonary Phthisis a Contagious Disease?" In considering the different examples which have been brought forward, it seems that the disease, when acquired by a person who had been previously healthy, is almost certainly the result of infection, and it is a remarkable fact that all the cases illustrating this fact have occurred in women, this being perhaps due to the relative constitutional debility and feebler power of resistance possessed by the female sex. In addition to this group of cases, the facts stated by Flindt in 1875, and by Reich in 1878, are of the greatest significance.

The following is a *resumé* of the former. In the autumn of 1872 an artisan, his wife, and five children (four sons between the ages of three and a half and fourteen, and a daughter fifteen years old) went to live in a village of Denmark, where they inhabited a small room already occupied by another family, consisting of a man, his wife, and a grown-up son affected by febrile phthisis (*phthisis florida*). They remained in this crowded and confined abode, in which the air was rendered truly poisonous by the presence of the patient, until Jan. 3, 1873, when they left the place for a more healthy dwelling. Before this, however, at the time of Christmas, the five children, who had been previously in good health, and unaffected by scrofula, suffered from pulmonary disease, and the evolution of the complaint was so rapid that they all died after the respective periods of seven weeks, three months, three months and a half, and seven months. The daughter, whose age was fifteen years, and who had only remained during one day in the infected room, was attacked like her brothers, and it was her death which occurred at the end of three months and a half. The autopsy was made in one case, that of the youngest boy, who lived for seven months. Several caverns were found in the right lung, numerous foci of yellow infiltration in the spleen, and tubercular ulceration in the small intestine, while the mesenteric glands were swollen and in a condition of caseous degeneration.

This fact, which is but little known, is itself, in my opinion, a sufficient proof of the possible infection of pulmonary phthisis. Without doubt, as I should at once state, this mode of transmission acquired in the above case an exceptional degree of power owing to the unusual concentration of morbific elements, but it is equally certain that the disease would not have been transmitted, whatever the means may have been, had it not been transmissible in itself and owing to its own nature. Thus, if a reserve is made as to the secondary question of degree, the case reported by Flindt seems to me to contain unfortunately but too conclusive proof of the contagion of the disease.

A case reported by Reich, in 1878, is of a different kind, but not less convincing. At Neuenburg, a village with thirteen hundred inhabitants, two midwives exercised their profession, dividing in almost equal proportions the obstetric practice of the place. In the winter of 1874 one of them was affected by phthisis, and died in 1876 from the effects of the disease. Of the infants, whose birth she had attended between April 4, 1875, and May 10, 1876, ten died of tubercular meningitis between July 11, 1875, and Sept. 29, 1876. Not one of these ten children was under the influence of hereditary predisposition. In the practice of the other midwife during the same period not a single child was affected by any form of tubercular disease. This alone is a strange circumstance, but the following details complete the demonstration. The first midwife was in the habit of removing by aspiration the mucous products which obstruct the *primæ viæ* in newly born children, and when the slightest degree of asphyxia occurred, direct insufflation was practised. It should be added that tubercular meningitis is not endemic at Neuenburg, where, in fact, it but rarely occurs, and that during the period between 1866 and 1874, of ninety-two children who died in the first year of their life, Reich observed but two cases of this disease, while in 1877, of ten children who died at this age, he observed but one case, and that in an infant whose parents suffered from tuberculosis.

These facts require no comment, and, to complete the examination of the question, a group of numerous and important cases which relate to the infective action of milk drawn from cows affected by tuberculosis will now be considered. The experimental researches of Klebs upon this subject have already been mentioned; those of Gerlach, Bollinger, and Fleming point to the same conclusion; and observation seems to show that the results were obtained in pathological sequence since Foot in 1877, Kommercil and Lochmann in 1878, openly proclaimed the possible transmission of the disease by the milk of cows affected by tuberculosis. The latter writer also asserts that a distinguished veterinary surgeon of his country, Thesen von As, made numerous experiments with regard to this question which always gave positive results. Leube, whose opinion is quoted in the work by Foot, admits also the transmission of tuberculosis by the milk of cows affected by phthisis, believing that in such cases the lesion occurs primarily in the intestine. Whether the simple ingestion of such milk can produce infection, or, as Chauveau thinks, a tubercular lesion must exist, in such a case, upon the udder of the animal, is a collateral question which cannot be solved at the present date.

It is, however, certain that the subcutaneous or intravenous injection of this milk may be followed by infection in the same way as its ordinary ingestion. The experiments made by Semmer in 1875 upon pigs and sheep leave no doubt as to this point. These were of the number of eleven, of which five gave positive results. Semmer also investigated the effects produced by injection of blood taken from a cow affected by tuberculosis, and of nineteen experiments which he thus performed eleven were positive, that is to say, were followed by the development of tuberculosis, this effect being always more decided in the pig than in the sheep.

The transmission of tuberculosis from the milk of animals affected by phthisis is almost universally denied or unrecognized in France. This, in my opinion, is a mistake, since its occurrence has been established for ten years, being based upon positive experiments and facts, and admitted by the most competent observers and veterinary schools in all other countries of Europe. Undoubtedly the transmission is not constant, and some experiments, such as those of Schreiber, and many of those conducted by Semmer and others, have given negative results. This, however, does not disprove the fact; for is there any infectious disease which is invariably transmitted whenever the conditions of contagion present themselves? Assuredly none. Why, then, should tuberculosis be expected to be more infectious than other diseases? There is nothing mysterious or unintelligible in the inconstancy of the results. Difference in the organic condition of the exposed person; the fact already stated by Klebs at the conclusion of his experiments, namely, that the transmission of tuberculosis by artificial ingestion of milk can be resisted by a vigorous organism, would explain the result. Nor is there anything in this varying and individual influence of the organism which is special to tuberculosis. It is a simple application of that general law which I have termed the law of receptivity (*la loi de receptivité*), a primary law which governs the ætiology of all transmissible diseases, and which is always fulfilled whichever mode of transmission be considered. In order to produce a contagious disease, a specific poison is required. But this alone does not suffice; the assent, the disposition of the organism to be affected by it, are also necessary, and this admission of or resistance to the infective agent is what I designate by the term receptivity (*receptivité*).

In my opinion, therefore, the negative experiments and facts do not disprove the conclusions authorized by those of which the results were positive. These, by their number, their clearness, and the multiplicity of their origin, are of incontestable value. It may be said, if the small number of cases be considered in which up to this time transmission is known to have thus occurred, that tuberculosis is the least infectious of such diseases, and that in this malady, instead of what occurs in diseases which are typically infectious, transmission, far from usually explaining its development, is but an exceptional and contingent cause. More than this, however, cannot be allowed, and the writings to which allusion has already been made conclusively prove the possible transmission of the disease.

The different modes in which this transmission has already been shown to occur are, firstly, by inoculation, a fact which is usually experimental, but which may occur by means of an accidental wound; secondly, by the inhalation of air more or less charged with diffusible elements coming from the patient, whether products of expiration, particles of expectoration, or perhaps, though less certainly, the evaporating perspiration, this danger being greatest to those who sleep together at night, and especially in badly ventilated dwellings; and lastly, by ingestion of milk supplied from cows affected by tuberculosis.

This truth is interesting, not only as regards children, but also with respect to adults, since the treatment by milk occupies at the present time an important place in the management of numerous diseases, and specially in that of pulmonary phthisis. The knowledge of these facts shows that vigorous supervision of the animals which supply milk is necessary, and especially in the southern provinces, where they are known to be more exposed to tuberculosis on account of their confinement in a stable, and the absence of suitable pasturage and fresh fodder.

In order to complete the different modes of transmission, the ingestion of meat obtained from animals affected by tuberculosis should also be mentioned; but since the facts which prove such transmission are less convincing than those named above, any definite statement with respect to this mode of contagion seems to me at the present time premature.

These ætiological notions, as I remarked many years ago, impose fresh duties on the medical attendant as regards the surroundings of the patient. Not only must the patient himself receive treatment, but the persons who are habitually in connection with him should be preserved as far as possible from all danger of infection, and the rules of prudence must not be transgressed by the enthusiasm of devoted affection. A well-ventilated apartment, the spray of carbolic acid or benzoate of soda solution used twice daily, the greatest cleanliness, the permanent presence of a disinfecting liquid in vases which receive the expectoration, and the disinfection of any linen or bedding soiled by it, are the principal and most important means to be adopted. I would direct special attention to the use of pulverized fluids for the purpose of purifying the air, a practice the advantages of which I have myself frequently appreciated, not only as regards the surroundings of the patient, but also with respect to the patient himself, in the treatment of whom this is a useful auxiliary. Even these precautions, however, are not all that must be taken;

one other condition should be fulfilled, at times difficult to realize, but which should be imposed without hesitation by the medical attendant—I mean the separation of husband and wife. These should not be allowed to share the same bed, or even the same room. This rule should be also followed in the case of children, who ought on no account to occupy the same room as a person affected by phthisis, in however early a stage the disease may be.

It is evident, therefore, that the ætiological facts which have been mentioned should produce great modifications in medical practice, and their consequence, no less important with regard to the special question of prophylactic treatment being now considered, is as follows. Every person who has been exposed during a certain period to one of the injurious influences enumerated above should be looked upon as threatened by tuberculosis, this very fact constituting an indication of preventive treatment, which is the more urgent in proportion as the constitutional condition of the person predisposes more to the disease.

Such are the principal facts drawn from the ætiological conditions of phthisis which indicate the advantage of prophylactic treatment. Certain individual phenomena point to the same conclusion.

One of the first of these presumptive signs is the external appearance special to the tubercular diathesis, of which the following are the most notable particulars. The features which characterize it are specially marked in adolescents and young persons; these have a slim figure, the chest and neck being elongated and thin; the muscles, and specially those of the neck and chest, are but slightly developed. On the other hand, the hair and eyelashes are remarkable for their length; the teeth are often most handsome; the eyes vivid, brilliant, and animated; the skin, fine and of rosy tint, allows a venous network of blue colour to be seen by transparence; the extremities of the fingers, however, are often deformed, being flattened, having a square shape, or ending in a club-shaped enlargement (hippocratic fingers, *doigts hippocratiques*). The person who presents these peculiarities is impressionable, with exaggerated nervous excitability, and frequently suffers from palpitations, while the character is unsteady, and apt to be irritable. There is liability to bronchial catarrh from the slightest cause, an affection which may soon be cured, but often continues for a length of time, depressing the patient without apparent reason. Others are apt to be out of breath, remaining in an overhot place, or an animated conversation rendering their respiration short and difficult, or perhaps altering the tone of their voice. When this constitutional condition coincides with suspicious family antecedents, when the infancy of the patient has been tainted by any scrofulous affection, the external appearance has all the value of a premonitory sign. Tuberculosis is at hand, and the warning signal must not be disregarded; the danger should, if possible, be averted by means of prophylactic treatment prudently carried out. Similar indications are furnished by the existence of anæmia, which resists all appropriate medication, of dyspepsia which is unaffected by suitable treatment, of menstrual disorders, of an alteration in the voice without appreciable cause, or of the frequent existence of vague, so-called rheumatoid pains in the walls of the chest or upper extremities.

In addition to these phenomena, of which the premonitory value has now become classical, certain other particulars should be mentioned which are less known, and which, without having so absolute a signification, should still be enquired about and taken into consideration. It is known that in healthy and well-formed persons the acromial is notably higher in position than the sternal extremity of the clavicle, so that the entire bone has an oblique direction from without inwards, and from above downwards. The observations of Aufrecht,

Hanisch, as well as my own, show that in those affected by phthisis the oblique direction from above downwards is replaced by one which is sensibly horizontal, so that the acromial and sternal extremities of the bone are almost at the same height. This anomaly is designated the deep position of the clavicle (*position profonde de la clavicule*). It is not a result of the development of tuberculosis but an original mode of conformation, and consequently when found to exist in a person who presents as yet no sign of a tubercular lesion, may raise suspicion and acquire the value of a premonitory indication. Further observation will determine how constant the relation is between this faulty conformation and pulmonary tuberculosis, but even now it may be regarded as a useful element in the appreciation of that question the solution of which we are now considering.

More important still is the phenomenon indicated by Waldenburg, which consists in the persistent diminution of inspiratory force appreciated by means of his pneumatic apparatus. In his opinion this is a symptom of tuberculosis at its development, but according to my observations is of value at an earlier date, perhaps preceding for some length of time the anatomical realization of the disease, and having all the importance of a premonitory sign. I differ also from Waldenburg in the mode of appreciating this symptom. He compares the inspiratory force of the person examined with the number expressing the mean force which would be exerted by a healthy person. This proceeding is obviously defective, as Fischl truly observed, since the number which is taken as the standard of comparison has but a mean value, and gives no information as to the inspiratory force exerted by the individual patient when in good health. It might well be that at that time the number expressing the inspiratory force of the patient was less than the mean used, while it still expressed the normal condition of the patient. In order to avoid this cause of uncertainty and error, the two terms of comparison should be taken from the same person, in which case the conclusion may be accepted without reserve. This object is gained by measuring the relation between the inspiratory and expiratory force, a relation which is constant when the person is in good health. Should this be altered, and the alteration which exists indicate a diminution of the inspiratory force, it is certain that such a change has really occurred; and if observations repeated at short intervals show that this anomaly persists while an examination of the respiratory organs proves them to be in a normal condition, the inspiratory weakness should be looked upon as a premonitory sign of pulmonary tuberculosis. In this comparative appreciation I make use of the data given by Brünniche in the following way. An adult in good health depresses the cylinder of Waldenburg 19·7 inches (50 centimetres) in ten inspirations, and elevates it to the same extent in seven expirations. This is the same as saying in simpler language that the inspiratory is equal to seven-tenths of the expiratory force. Upon this basis, if one examines the combined effect of a series of ten inspirations, or if one merely compares with each other the effect of one inspiration and expiration, an exact notion of the relative force of inspiration in the person examined may always be obtained. I strongly recommend such measurement in suspicious cases, which furnishes, without much trouble being taken, information of great practical value as regards the anticipatory diagnosis and prophylactic treatment of the disease.

In the conditions which we are now studying, inspiratory weakness is due to more or less complete inaction of the upper part of the lung, a fact which explains its premonitory significance. Such inaction, in fact, is recognized as the condition which is most favourable to the development of tubercles, and as the cause of the lesion being first localized at the apex of the lung. The researches recently undertaken by Hänisch, by means of a stethographic apparatus of his own invention which registers the inspiratory movements in corresponding portions of the chest, have furnished new and precise confirmation of these ideas.

Amongst the signs of imminent tuberculosis, Aufrecht has mentioned the persistent or repeated development of pityriasis versicolor; this fact is stated with the view of completing the account of these indications, and owing to the good authority of the observer, though I myself have had no personal experience of its occurrence.

The premonitory value of the different phenomena which we have been considering is real so far as each of them is separately concerned, but is far greater when a certain number of them are combined in the same person. Without doubt the practical benefit derived from these premonitory indications is often lost, because the adoption of medical treatment occurs too late in the complaint, being applied not to a possible but to a confirmed disease. This fact, however, in no way diminishes the value of the premonitory indications. While there is still time they should be sought with persevering care, and if this course is followed, more often than would be supposed, the treatment can be begun at so early a date as to deserve the name of prophylactic. It has then much more chance of success; for, as I have already said, and as I repeat at the end of these considerations, the preventive has certainly more power than the curative form of treatment as regards the final issue of the disease.

CHAPTER V.

PROPHYLACTIC TREATMENT (*Termination*).

Object of prophylactic treatment; its correspondence with the ætiological indications.

Ætiological indication drawn from constitutional debility—Means of prevention and cure—Advantage of the latter—Causes of failure.

Hygienic measures adopted in prophylactic treatment—Residence in the country — Diet — Hydropathy—Gymnastic exercises—Selection of residence—Sea voyages.

Marriage of those who are predisposed to phthisis.

Ætiological indication drawn from pulmonary insufficiency—Aërotherapeutics —Practice and effects—Comparison between the effects of expiration in air rarefied by means of an apparatus, or by the height of the residence.

Medication in prophylactic treatment—Indications furnished by anæmia, or scrofula; by congestion or inflammation of the respiratory organs.

WITH the exception of those rare cases in which the disease is due to contagion, and in which also, as should not be forgotten, the question of transmission depends above all upon the organic strength of the individual concerned, it is constitutional debility, that is, the so-called hypotrophy, which produces the morbid liability to tuberculosis. The object and end of prophylactic treatment should, therefore, be to modify the nutritive conditions of the patient, and to transform a soil which was favourable to the development of tubercles into one which refuses to bear them.

Besides this fundamental fact, connected with the ætiology of the disease and furnished by the general condition of the patient, a special indication of no less importance, drawn from the relation shown empirically to exist between the inaction of the lungs and the development of tubercle, should be taken into consideration. That tubercle has a special affinity for organs whose functions are incompletely performed, is a fact of absolute certainty, and the predilection which it shows for the apex of the lungs has no other cause than the comparative functional inaction of that part. This difference between the apex and other parts of the pulmonary organs is a constant, normal, and physiological fact; but whilst in the healthy this difference is not so great as to be incompatible with the normal structure of the part, in those who are affected with constitutional debility it exceeds this limit, and exists to an extent corresponding with their debility. Thus in such persons the whole of the lungs, like other organs, performs its functions subnormally, and this diminution of activity is the more pronounced in proportion as the general debility is greater. Since the apices are normally inferior to other organs in this respect, a condition of absolute inaction may be then reached in these parts, which is eminently favourable to the formation of tubercles. Such facts explain the assertion made above as to the important indication furnished by inaction of the lungs.

Owing to the fact that this bears a constant and proportional relation to existing malnutrition, it naturally diminishes when the general state of the patient improves, and is thus indirectly controlled by treatment applied to the constitutional condition of the patient. This secondary action should not, however, be considered sufficient in all cases, and very frequently it is more prudent to meet this important indication in a direct manner, and with such special treatment as has the effect of maintaining at its maximum the functional activity of the lungs.

Such are the constant and uniform fundamental indications of prophylactic treatment. There are, in fact, two ætiological indications, one drawn from the general condition of malnutrition, the other from the functional inactivity of the lungs, notably at their upper part. There may possibly be other indications from the existence of some diathesis in the individual or family, but in all cases these two considerations exceed others in importance.

Setting aside for the present the relative question of pulmonary inaction—which will be considered before long in full detail—a clear and precise idea may be obtained of the task imposed upon preventive treatment. Its object is to produce a change of constitution, and the treatment to be imposed may be said to include the whole physical training, and to depend as much, if not more, on hygienic as on medical agents.

Upon this point the opinion of medical authorities is still divided. Some wish to gain their object by removing those who are predisposed to the complaint from every external influence which might favour its development. Dreading, and not without good reason, the effects of bronchitis, the person is withdrawn from exposure to this illness by means of rigid confinement, and careful precautions against whatever might produce a chill. The idea of others, who follow the example of Graves, and have more extended and legitimate views, is to fortify the patient, enable the constitution to resist these deleterious influences, and successfully meet whatever indisposition or disease might be produced by the effects of cold.

The first plan removes injurious influences, the second enables the patient to resist them; the first plan protects, while the second inures. To protect, or to fortify, are the two tendencies which are not irreconcilable with each other.

In my opinion, the latter plan of fortifying is undoubtedly the best. A moment's reflection will show why. The means adopted in it are those which in all conditions are most adapted to strengthen the constitution by ensuring the integrity of both the general and local nutrition. Thus, both in its object and in the means which it necessarily adopts to attain that end, this method necessarily agrees with the ætiological indication which results from constitutional malnutrition. The other method, on the contrary, viz. that of protection, if fully employed, is carried out—how? By accumulating together every imaginable precaution against the action of air, or the accidental possibility of a chill; by permanent confinement to the house, at any rate during the cold season; by the avoidance of bodily exercise, and by maintaining as uniform a temperature as possible, which soon becomes from habit so necessary that it cannot be omitted without danger to the patient. Such are the means adopted; means which necessarily incapacitate the person, rendering him unable to resist injurious influences, favouring instead of opposing constitutional debility, and possibly themselves producing, in the absence of existing predisposition, a condition suitable to the development of phthisis. The tuberculosis observed in confined animals proves the accuracy of this statement. The pretended method of protection, which really incapacitates the person, is, therefore, directly opposed to the object indicated, and since the commencement of my practice I have always strenuously opposed this deceptive plan of treatment, which in my opinion should be condemned and unreservedly abandoned.

Prophylactic treatment, based upon the plan of fortifying, the advantage of which I asserted in 1862, has, from a theoretical point of view, almost unlimited power. Numerous, however, are the disappointments in its practical employment—disappointments resulting from two causes, which are totally independent of the inherent power of the method. On the one hand, medical intervention is most often applied too late, either from neglect on the part of the patient or his family, or because the medical attendant, ill informed as to the value of the premonitory signs of tuberculosis, considers himself authorized to pursue an expectant plan of treatment so long as he can certify the absence of any actual lesion. The cause, indeed, is of little consequence, but the fact is certain that in a large number of cases the premonitory period is not utilized at all, or so late that the so-called prophylactic treatment is really confused with that of the confirmed disease.

On the other hand, preventive treatment, as I understand it, comprises among the number of its most useful and indispensable agents many hygienic measures which cannot easily be reconciled with the family habits, the requirements of intellectual education, or, in short, with the ordinary routine of life. Too often, again, all medical advice is disregarded, legitimate requirements are looked upon as exaggerated pretensions, the evil to be opposed is regarded as but a remote possibility, the danger to be avoided as the idea of a pessimist. In such cases, as may well be conceived, it is by no means easy to induce those interested to submit to such hygienic reform as necessitates a total change in life, and upsets all the habitual conditions of existence. The result of this is a series of arrangements and plans discussed at great length, and ending in the adoption of such of the means proposed as are agreeable and easily carried out, whilst those which are considered inconvenient or ill timed are disregarded. Hence results a treatment by half-measures which is almost certain to fail, the failure being due, not to the powerlessness of preventive treatment, but to its incomplete adoption.

In other cases which are most deserving, and at the same time most sad, an absence of resource prevents the family from complying with medical advice. What is possible can then alone be done, and want of success proves nothing as to the value of this treatment, which in reality has never been put into practice.

All these difficulties in medical practice should be known; they are constantly occurring, not only in the prophylactic, but also in the curative treatment of pulmonary phthisis. In the midst of such obstacles, which repeatedly present themselves, the medical advice should be clear, and whatever seems most to the interest of the patient should be insisted upon. If, when this is done, there is a real or supposed impossibility to carry it out, the medical attendant will have done his duty, and have preserved his responsibility from all criticism.

Having said this, the measures will first be mentioned which are in my opinion indispensable, that the prophylactic treatment may be completely carried out, and it will then be seen that the difficulties which exist in their practical application have not been exaggerated.

In a group of cases unfortunately but too large, the prophylactic treatment should be commenced in infancy, in fact from the time of birth; this should be done in children belonging to families in which the hereditary transmission of the disease is to be feared. When such inheritance would be direct, that is to say, when tuberculosis exists in one of the parents, the mother should be absolutely forbidden to nurse her child. It might be supposed that this precaution is less necessary when the father alone is affected by the disease; such, however, is not my opinion, and in that case also lactation should be performed by another person. The mother, it may be truly said, is healthy, but from constant association with her husband she is always in danger of contracting the disease, so that at any moment the milk may become an efficient agent in transmitting the disease from her to the child. Thus it is more prudent to act in such cases as if this possibility was certain to happen, and whenever direct hereditary transmission is possible from either parent being affected by the disease, the mother should be forbidden to suckle the child, and a healthy and vigorous nurse should be put in her place. Lactation should then be continued for as long a time as possible; that is to say, for eighteen, twenty, or two and twenty months.* The child should not be weaned suddenly and totally, but the human milk be succeeded by a mixed diet, of which ass's or goat's milk forms a decided part. Nor is the management of lactation the only question to be solved in the condition which we are now considering. Other measures are no less indispensable, and their necessity is so evident that it seems almost useless to state them, though it is perhaps to these that the greatest opposition will be offered. Under no pretext should a child be allowed to sleep in the same room as the affected parent, or to enter it until it has been well ventilated, nor is it prudent for him to remain there for any length of time. Parents should also be careful, when kissing their children, not to bring their lips in contact with those of the child, as they so frequently do when the latter is an infant.

When collateral hereditary influence is alone to be feared, the mother may be allowed to nurse her child, so long as this is not counter-indicated by the condition of her health, which is frequently the case, as will soon be seen. With this difference, the rules which govern the physical education of children apply not only to the two forms of hereditary transmission, but also in a general way to children in whom it is desired to develop constitutional vigour and power of resistance.

The children should be brought up in the country, and remain in the open air as many hours as possible during the day. Their clothing should be short in length, and produce no constriction either at the waist or upon the chest. Baths should be taken daily, tepid in winter, but cold during the rest of the year. The baths should be short in duration, and followed by forcible dry friction. Thus healthful stimulation

* Such a prolonged period of lactation is not recommended by English authors.

of the functions of the skin is produced, the power of resisting the impression of cold is increased, and preparation is made for treatment by hydropathy, properly so called, which is one of the most powerful agents in prophylactic treatment. The muscular development must be watched with the same care, and daily exercise in walking and gymnastics, in which the duration and strength required increase proportionately to the age, should be ensured. When the child has reached such an age as to take ordinary diet, the food should be of a substantial character, being chiefly composed of roasted or broiled meat, which is underdone. Wine should be taken from the first in quantity proportional to the age and excitability of the child. If, when such measures are properly carried out, the child still seems insufficiently nourished, the constitution does not gain strength, and the heat of the body is badly preserved, medication of a tonic character should be instituted, by means of syrup containing iodide of iron or cinchona, and such hydro-carbons as are easily combustible, of which cod-liver oil is the type. In these children no kind of indisposition should be neglected. Intestinal troubles, above all, should not be allowed to become chronic, aggravating as they do the debility which it is wished to check. Mental study should be deferred until a later date, and lessons must be begun with great care and be given at home. This question will shortly be again considered.

Such hygienic precepts, to which the most serious attention should be paid, concern the time of infancy. They form, as it were, the first stage in prophylactic treatment in those unfortunately but too rare cases in which it can be completely carried out, that is from the time of birth. The rules which will now be mentioned apply to all ages and circumstances. I believe them to be equally necessary for the success of the treatment, but unfortunately the most important are precisely those of which the application meets with the most serious obstacles.

The first condition required to maintain and increase the organic strength is that the air breathed should be continually renewed. The great importance of such *pabulum vitæ* is a fact known from ancient times, and it is needless to dwell upon it. Hence I lay down without hesitation the following rule. A person in whom, for any reason whatever, the development of tubercle is considered probable, should live in the country so long as any suspicious constitutional debility remains, and until the age which is specially to be feared has been passed without disease. Such a residence should be permanently occupied both in winter and summer, the air in towns being less favourable than that of the country even in the cold seasons; and the dangers of a chill, when once the method of confinement is abandoned, are not greater in the country than in town. A country life, then, has certain advantages, and furnishes the only means of fulfilling a fundamental part of the prophylactic treatment, namely, that of living in air which is pure and constantly renewed.

It need scarcely be said that the hygienic rules which relate to the dwelling occupied should be rigorously followed. This should be dry, and situated on ground which is higher than that of the surrounding regions. The person whose health is considered should not inhabit the ground floor, and the rooms occupied by him should certainly look towards the south. His bedroom should only be occupied at night, and by day should be well ventilated, this being prevented by rainy weather alone. In winter the temperature of the bedroom should not exceed from 54° to 60° F. (12° to 15° C.) at bedtime. I cannot recommend the American rule of opening the window at night, even in cold weather, though in summer it is advisable to have the window open in the adjoining room, with the door which leads into it widely thrown back. One thing is also necessary, namely, to avoid exposing the bed to a draught of air coming from outside the house.

In all seasons of the year the patient, or rather the person whom it is wished to preserve from illness, should wear flannel over the body and upper extremities. He should be warmly clothed, though not in excess, and even in summer should be careful to put on extra clothing in the early morning. At the same time, the use of thick scarfs round the neck, comforters or hoods which surround the head, should be absolutely forbidden—a bad habit, which is a constant cause of sore throat, laryngitis, and tracheo-bronchitis. The food taken should be of a substantial character, as has just been mentioned; but it is better to avoid ceaseless monotony in what is taken, which cannot fail to produce a distaste for food and loss of appetite. So long as the digestive functions are in a healthy condition, a small amount of fatty and farinaceous food may be introduced into the diet. The time for meals should be arranged in such a way that at least two hours and a half pass between the end of the last meal and bedtime, and it is also better that this meal should be less abundant than that taken at midday. I also prescribe in all such cases that warm milk should be drunk at the moment of its being drawn from the cow to the extent of two cups at least daily; one in the morning fasting immediately after the shower-bath of which I shall speak, the other in the afternoon at a suitable time before supper. The taste of the patient need alone be considered as to the choice between milk from the cow, goat, or ass. At the same time, the milk from the two latter is less often taken for a length of time without distaste, a consideration which is by no means without importance. Should cow's milk be chosen, the precautions shown to be necessary, by the facts relating to transmission of tuberculosis by means of milk having this source, must not be forgotten. The persons interested should be informed how necessary it is that a constant watch should be kept over the hygienic conditions of the animals, specially as regards the cleanliness of the cowshed, fresh fodder forming part of the food, and their life being spent as much in the open air as the season permits.

Even when these fundamental rules are laid down as regards dwelling, clothing, and diet, the duty of the medical attendant is not yet done, and the way in which the patient's life is to be spent daily should be arranged with the same care. I do not insist upon his spending a long time in bed, which should not exceed seven or eight hours at most. Immediately after rising, which he should do at an early hour, a cold shower-bath should be taken in the form of rain, accompanied by a larger jet of water. This should be continued for a very short time at first, namely, for twenty or thirty seconds, the time being gradually prolonged to a minute, or rather more, according to the mode in which the reaction occurs. The shower-bath is to be immediately followed by forcible and dry friction, and the patient, when dressed, after having drunk one or two cups of milk, should take a walk, uninterrupted but not long enough to produce a feeling of fatigue. This practice of hydrotherapeutics is one of the most powerful means of re-establishing that organic strength which is the object of all such efforts. It also constitutes the best preservative against the dangers of a chill, and maintains a healthy activity in the functions of the skin, which is indispensable, that nutrition and hæmatosis may be properly carried on. Nor is this all; the cold shower-bath properly used may become the cause of true pulmonary exercise, and in this way answer to the indication furnished by inaction of the lungs.

The shock which is occasioned by the first contact with water causes respiration to cease; but after the first instant a reaction occurs, the breathing becomes deep, and the patient acquires the habit of exaggerating it by the use of all his respiratory power; when once this habit is acquired, the shower-bath is not only a general stimulant, but also a cause of pulmonary activity and expansion, and thus fulfils, at least to some extent, the two fundamental objects of prophylactic treatment.

If the absence of the proper apparatus makes it impossible to use the shower-bath, the want must be supplied by general lotions applied every morning, and composed of cold water mixed with vinegar or alcohol; in this way the same good effect on the cutaneous functions may be obtained, though the general tonic action is much weakened, and the effect upon pulmonary expansion almost entirely lost.

It is needless to say that with delicate and sensitive persons, who, though still in the premonitory period, evince that nervous excitability of which I showed the importance, it is necessary to proceed by gradual changes. The shower-bath or cold lotion should not be prescribed at once for a young man or girl who has never been accustomed to its use. At first the patient should be bathed with tepid lotions or sprinkled with tepid water, and the cold hydrotherapeutic treatment should be gradually adopted, the persistence as well as the rapidity of the reaction being taken as a guide in this matter. The persistence of reaction is the true criterion. It is not uncommon, specially in those excitable persons whom I have mentioned, to observe a temporary reaction, as to the importance of which it is necessary to be enlightened. After the friction which follows the shower-bath, or in any case after a few minutes' exercise, there is a satisfactory and complete reaction; this, however, only lasts during the walk, and when the patient is at rest it ceases, and gives place to a sensation of cold which is often most distressing. The organic action provoked by the cold water evidently surpasses the powers of the patient. After a temporary effort, which is in such cases a mere expenditure of strength, the temperature falls to below its normal condition, and if this sign is overlooked, the shower-bath becomes a depressing agent, and the nervous excitability rapidly increases. In cases of this kind, the advance should be made with care and gradual change, and it is then likely that, though many delays may occur, the complete method of treatment will be ultimately developed which is one of the most important elements of success.

In the second half of the day some hours should be again passed in the open air, and when the walk is over, the time will be reached which is most opportune for the afternoon glass of milk. Walking, however, is not the only form of exercise which should be taken; if the character of the country is suitable, constant ascents, proportionate to the age and strength of the patient, should be prescribed. These ascents should be made with slow and measured steps, so as to occasion no fatigue to the respiratory organs, and there should be occasional rests on the way. If it is wished to expand the lungs as far as can possibly be done, the person should be advised while climbing to place a stick between the arms, which are thrown back, and the dorsal region of the back. In this position the transverse diameter of the lower part of the chest is completely expanded, the fixed position of the upper limbs causes the whole action of the auxiliary inspiratory muscles to be combined in raising the chest, and the upper part of the lungs, whose expansion is always measured by that of the chest, dilates as far as possible at each inspiration, which is necessarily exaggerated by the effect of the ascent. When carried on in this way, such a plan constitutes a true and useful kind of pulmonary exercise.

I should also mention riding as a means of inducing physical development in the conditions which we are now studying; at the same time, I do not consider it absolutely necessary in the treatment, and in its place recommend that in every case gymnastic exercises should be practised daily. I insist especially on those which are performed by means of the thoraco-brachial muscles; as, for example, the methodical use of weights coupled together, and known as expanders.

When walks and out-of-door life are prevented by bad weather, such practices are an indispensable resource for the exercise of muscular activity.

The kind of life of which I have been tracing the rules is not incompatible with intellectual education, as it leaves a certain number of hours unoccupied in the morning and afternoon; these can be employed in study, which, however, should only be carried on, specially at first, to the extent justified by the constitutional condition of the patient. The selection of a country residence is not unimportant, being, on the contrary, one of the chief elements of success. The person should reside in a mountain climate even in winter, and if this is impossible, and the place where he ordinarily lives at a low elevation, he should spend at least several months of the year in mountain air. I will not now discuss the subject of climatic stations, for, as I said before, I wish to reserve the examination of their relative advantages until a later chapter, and limit myself for the present to the general statement that elevated or mountain climates alone are suitable during the period of prophylactic treatment. When family circumstances do not permit the patient to live in the country, the most important means of treatment cannot be employed, but rigorous application of the other measures, namely, out-of-door life, gymnastic exercises, hydrotherapeutic treatment, etc., should still be enforced, and every means should be adopted to obtain for the patient, at least during some months of the year, the benefit of pure country air. Prudence should be exercised in the choice of studies, and on no account should any confinement in colleges or schools be allowed, where the close atmosphere of the schoolroom and dormitory exerts an injurious influence by day and night.

It might be, and in fact it has been, doubted whether sea voyages are of service in cases of confirmed phthisis; but in the prophylactic treatment of the disease, specially in cases where the invalid cannot habitually reside in the mountains, their advantage is, in my opinion, incontestable, so long as the voyage fulfils one condition which I consider indispensable, namely, that of being sufficiently long, and I will explain the reason. Sea-sickness is a cause of great fatigue even to a person in the best of health, and experience shows that one who is specially susceptible does not become accustomed to the sea until at least five or six days have passed. This means that the first few days of the voyage are devoted to sea-sickness, so that if the voyage only occupies twelve days, for instance, it may be assumed that the immunity of the second part will hardly compensate for the fatigue caused by the suffering of the first; as for any benefit, there would in such a case be none. If, on the contrary, the voyage lasts two months, it is almost certain that a period of six weeks will exist during which the traveller will derive real and lasting advantage from the strong tonic influence of the sea air. Under these conditions a sea voyage, although in many ways so anti-hygienic, is an efficient means of fortifying the constitution, and may be recommended with confidence, though only during the prophylactic period which is now under consideration.

Before ceasing to consider the hygienic measures which should be adopted in the prophylactic treatment of phthisis, a question of some delicacy must be discussed. Should persons affected by the debility which precedes tuberculosis be permitted to marry? If no preventive treatment has been adopted, or if such treatment has failed to produce the desired constitutional change, most assuredly not. The effect of marriage may be to transform a virtual into a real disease, and the children born may suffer in consequence of this condition. I have seen many conclusive examples of both these facts. Marriage should also be forbidden, even though prophylactic treatment may have produced a satisfactory result, if the suspicious predisposition is of hereditary origin.

On the other hand, when such treatment has produced at the adult age a complete constitutional change, marriage may be permitted, without reservation as far as the male sex is concerned, but not without some doubt in the female, on account of the possible exhaustion produced by repeated pregnancy. This is a question which regards the individual condition of the person, and does not admit of a general answer. In any case, when such a marriage occurs, the mother should not be allowed to nurse her children. I have already seen three cases in which phthisis was developed as the consequence of prolonged lactation (from eighteen to twenty-two months), hereditary predisposition being totally absent, and without other possible cause than the debility produced thereby. The first of these relates to a young woman of Saint Maur, whom I saw in conjunction with my distinguished colleagues, Dr. Bacchi and Dr. Tourasse, and in whom the tuberculosis made rapid progress. This patient had suckled her child for twenty-two months. The knowledge furnished by these facts should not be forgotten.

The preventive therapeutic treatment of which I have just described the hygienic details meets by its various methods the requirements indicated by the existing malnutrition; nay, more—by some of them, such as the shower-bath, gymnastic exercise, and methodical ascents, it also fulfils the special necessity which results from inactivity of the lungs. These indirect means, however, fulfil but imperfectly this important indication, and in many cases cannot be employed without failure. That the treatment may be complete and efficient, it is then necessary to resort to more direct agents; hence arises the question of treatment by compressed air.

After trying for a certain number of times the effect of forced inspiration, and expiration produced in the patient twice daily by the maximum action of muscular power, it will be recognized that this method, which is so good from a theoretical point of view, and which may, in fact, be of some benefit, is defective and insufficient to attain the desired object. It is defective because the exercises, which are irksome and distressing to the patient, cause considerable fatigue, and insufficient because the energy expended in performing the respiratory movements is necessarily subordinate to the muscular strength of the patient. In those persons in whom the question of prophylactic treatment arises, the muscular force shares in the general debility, and is incapable of producing complete expansion of the pulmonary organs. This being the case, it is easy to understand the great advantage of a method which really enables the desired result to be obtained; that is to say, the maximum expansion of the lungs to take place without fatigue to the patient, and independently of the muscular strength which he has at his disposal. This method does exist, and the means of applying it are supplied by the pneumatic apparatus whose employment constitutes the treatment by compressed air.

The general principle is as follows: to obtain by change of atmospheric pressure in the air breathed the same effect as by voluntary or forced contractions of the respiratory muscles; in other words, to modify the atmospheric pressure in such a way that the mere act of breathing naturally in an artificial atmosphere produces the same increase of pulmonary expansion.

This principle was first carried out by increasing the pressure of the air breathed; that is, by causing respiration to take place in air compressed until its tension exceeds the normal atmospheric pressure by a fraction varying from $\frac{2}{10}$ to $\frac{5}{10}$. Two kinds of apparatus are formed upon this plan. In one the construction is fixed, and composed of pneumatic chambers having no free communication with the external atmosphere, and in which the air is caused to exert an exaggerated amount of pressure which is measured by means of a manometer. The treatment consists in the patient remaining each day for an hour, an hour and a half, or two hours, in such a chamber, entirely surrounded by air which exerts this pressure and which he also breathes. The other kinds of apparatus are movable, being termed on that account "transportable." In these the air is compressed in a cylinder, from which a tube branches off provided with a mouthpiece, by means of which the air of the cylinder alone can be taken into the lungs during inspiration, whilst expiration takes place in free air. In both cases the effect is independent of the muscular energy expended. Whether the pneumatic chamber or cylinder is used, there is no need that the breathing should be laborious, and ordinary respiration of the usual force and frequency is alone needed. The increased pressure of the air mechanically and of itself produces the supplementary expansion, which not long since could only be obtained by the problematic and inconstant effect of muscular action.

The observers who have studied the effects of breathing compressed air are not unanimous as to the different effects which this treatment has upon the circulatory system; an increase of pressure in the aortic system is unanimously admitted by Waldenburg, P. Bert, Jacobson, Lazarus, Neukomm, and others, but its more remote effects upon the systemic or pulmonary circulation have been differently interpreted.

Whilst Waldenburg admits that increased inspiratory pressure should interfere with the flow of blood from the large veins into the right side of the heart, and thus produce a condition of plethora in the systemic, and a proportionate diminution of the quantity of blood in the pulmonary circulation, Dührssen, on the contrary, believes, and in my opinion not without good reason, that since increased pressure in the aortic system promotes the general venous circulation, it would effectively compensate thus for the obstacle presented by the increased inspiratory pressure to the flow of venous blood into the right side of the heart, so that the cardio-pulmonary would preserve its normal relation to the aortic circulation. Besides this—and Schnitzler has already made the same remark—the theoretical deductions of Waldenburg are not confirmed by the therapeutic results obtained by means of the treatment by aërotherapeutics (*l'aerotherapie*). If such deductions were true, this method should at least be as beneficial in diseases of the heart as in those of the respiratory organs. The results of observation, however, so far at least, are opposed to this theory.

I will not dwell longer upon these secondary questions, which would lead us far from the subject which we have now under consideration. The agreement, on the other hand, is general as to the relative effects of activity in the respiratory functions, a fact about which no uncertainty should be felt.

The meaning of the term respiratory or vital capacity is well known. It implies the quantity of air which can be drawn into or driven out of the lungs by the alternating movements of inspiration and expiration. When the movements are performed with the highest degree of energy, this capacity is equal to 193·6[*] cubic inches (three litres and a half), but in the normal respiration of the healthy adult the mean quantity does not usually exceed 30·5 cubic inches (half a litre). This capacity can be exactly measured by means of more than one apparatus, known as a spirometer, such, for example, as that of Hutchinson, or of Schnepf, or, again, by the anapnograph of Bergen and Kastus. The first effect of breathing compressed air is increase in the pulmonary capacity. According to the observations of Liebig, such increase is but slight when each day is considered separately, but after a certain time, varying in length from twenty-five to forty days, this increase may be as much as one-fifth of the primary capacity. This fact is revealed, without any

---

[*] As Dr. Hutchinson explains ("Trans. of the Royal Medico-Chir. Soc.," vol. xxix.), it varies according to the height, weight, and age, this being the average vital capacity of a healthy person 5 feet 8 inches in height at 60° F.

special measurement, by the increased amplitude of the inspirations in compressed air, which, according to Mosso, sensibly diminishes when a return is made to the normal pressure.

When this form of treatment is begun, the increase in pulmonary capacity and expansion, which necessarily correspond to each other, is but temporary, and does not persist when the natural air is breathed. If, however, this plan is continued for a sufficient length of time, as in the mean for thirty days, these favourable changes continue to exist, the vital capacity is definitely increased, and the expansion of the lungs in a state of rest exceeds the limits which previously existed. Hence it results that the inspiration of compressed air not only constitutes a salutary process capable of increasing the pulmonary expansion whilst it is continued, but may also produce a permanent change. This fact, affirmed by all observers, and which I myself have often proved to be true, gives to this mode of treatment an incontestable superiority in all cases when it is advantageous to increase to its maximum, or to maintain at that point, the capacity and expansion of the lungs.

The researches of Stembo and Schirmunsky, made under the direction of Lazarus, have shown that the inspiration of compressed air causes more complete opening of the air vesicles, with subsequent retraction corresponding to the increased call previously made upon the elasticity of the tissue. This might have been foreseen, since it must necessarily accompany the increase in vital capacity which has been proved to occur. The renewal of air in the respiratory organs is consequently more complete, and the hæmatosis more actively carried on. On the other hand, the intrathoracic inspiratory pressure, which in the normal condition is negative in character, that is to say less than the external pressure, is increased by the influence of compressed air; and the observations of Waldenburg show that if compressed air is breathed such pressure may in time become positive. This condition we have seen to be of such a character as to produce more activity in the cardio-pulmonary circulation.*

Thus, in short, the repeated inspiration of compressed air has the following mechanical effects: temporary and then permanent increase of the vital capacity, increase of pulmonary ventilation, increased expansion or unfolding of the lungs, increase of the intrathoracic inspiratory pressure, and of activity in the pulmonary circulation. Thus it is obvious that this form of treatment is admirably suited to the group of cases which we are now considering; it causes the upper parts of the lungs to act; it promotes ventilation and circulation in those parts, and thus prevents the direct consequences of inaction, which is so favourable to the formation of morbid products of a low order, such as tubercle. The best preservative against such formations is activity in the circulation and functions of the lung, and this is carried to an extreme degree by the above method.

Whenever, therefore, respiratory inaction is found to exist, specially in the upper part of the lung, in a person who presents other indications that the prophylactic treatment of phthisis would be advisable, aërotherapeutic treatment should be employed. A fixed apparatus should be used if, in opposition to medical advice, the patient resides in a town of such size that establishments exist therein which enable this to be done; a transportable one if he resides, as he should do, in the country. There is no necessity in ordinary practice to employ the complicated machines already mentioned for the purpose of recognizing such respiratory inaction as would indicate that this treatment is suitable. It is sufficient to examine comparatively and with care how the functions of the lung are performed at the upper and middle part, and there should be no hesitation in making use of the general information which experience has furnished. Pulmonary inaction may thus be

* See pp. 104, 105.

presumed to exist in all persons who have a slim chest, which is thin and narrow, horizontally placed clavicles, and a slightly developed muscular system, whatever may be the other characters of the external appearance.

The advantages of aërotherapeutic treatment are not confined to its mechanical action. In order to appreciate its whole value, regard must be paid also to its chemical effects, which are the following: increase in the quantity of oxygen absorbed, and in the proportion of carbonic acid gas in the air of expiration. These modifications produce in time a more active state of nutrition, shown by increased weight of the body, and, as Liebig expresses it, by removal of those signs which are special to the phthisical condition. In consequence—and this is an important fact—the above method is beneficial, not only as regards the local indication drawn from the functional action of the respiratory organs, but also from its effectively co-operating in the fulfilment of those general and fundamental indications which are furnished by the constitutional malnutrition.

These are certainly advantages of the greatest value, and, persuaded of their reality, I can only regret that they are not more often utilized. With regard to the effects, which are purely mechanical, what is even better may be done, as will soon be evident.

In the employment of aërotherapeutic treatment by means of compressed air, the act of inspiration is alone modified, expiration being totally unaffected. Thus, when the transportable apparatus is used, the latter takes place in the open air, while in the pneumatic chamber it occurs in compressed air, so that if any change at all is produced, it must consist in retardation due to the obstacle presented to the expired air by the medium which receives it. The expiratory activity remaining the same, the conditions which affect the air remaining in the lungs, the so-called residual air, or reserve supply, remains the same, or at most is but very slightly diminished in quantity, owing to increased retraction of the lung when expiration occurs. Such retraction, as I have previously explained, is proportional to the increased call made upon the elasticity of the tissue by the inspiratory expansion. In addition to this modification, which is interesting, though small in extent, the expiration, and consequently the residual air in the lungs, does not present any change when aërotherapeutic treatment is practised by means of compressed air.

It can be conceived, however, that it would be useful to increase the force of the expiratory in the same way as of the inspiratory act, without the effect of muscular contraction being required. If this possibility could be attained, what would the consequences be? Expiratory retraction carried to the farthest point, and hence a new cause of increase in the activity and circulation of the lungs, diminution of the residual air to its least amount, and consequently as complete renewal as possible of the air in the lungs. In these conditions not only is respiration more active as regards the quantity of air breathed, but the quality of such air is better, since the fixed residuum which forms in reality a confined medium becomes notably reduced in amount. These statements are not suppositions but facts; in principle such changes occur when expiration takes place in rarefied air, in practice when by means of well-constructed machines the inspiration of compressed air is succeeded by expiration in a rarefied medium. The lung, expanded to the utmost extent by inspiration, is reduced in size, and as far as possible sets free the air which it contains by means of expiration. The reason of this is that the rarefaction of the medium which receives the expired air acts as it were by attraction on the pulmonary substance, removes from it all the air necessary for re-establishing an equilibrium of pressure, only leaving the small residue known in physiology by the name of residual air, that is to say, the amount which cannot be expelled by the most energetic expiration. In varying within limits which are compatible with the performance of normal physiological functions, the degree of rarefaction of the air in which expiration takes place, the power of attraction resulting from the absence of equilibrium and the quantity of air removed from the lung, may be made to vary in a corresponding degree.

This being what occurs when expiration takes place in rarefied air, it is evident that one absolute counter-indication exists to this practice, namely, the tendency to hæmoptysis. If this has ever occurred, to however slight an extent, sufficient indication exists that this treatment should not be adopted. If, however, this form of hæmorrhage has not actually taken place, and the tendency alone exists as shown by a suspicious constitutional condition associated with the cardio-vascular excitability already mentioned, it will not be actually necessary to abstain from this practice though it should not be adopted without great circumspection, and the pressure of the medium which receives the air expired should be gradually diminished by not more than a fraction of a cubic inch at a time. It is easy to understand that if the rarefaction of the air be suddenly and notably increased, the attractive force which results from absence of equilibrium, and which mechanically produces dilatation of the blood-vessels, may determine their rupture should such dilatation be excessive in character, or even in cases of slight dilatation when that abnormal weakness exists which so often accompanies the constitutional conditions in question, and which, in short, is the organic stratum upon which the hæmorrhagic tendency is based.

One fact should be mentioned which has not yet been done, as far as I know, and which is by no means without interest. There is a special reason why expiration in rarefied air, such as occurs in the pneumatic machines, should expose the person to bronchio-pulmonary hæmorrhage, namely, the maintenance of the normal atmospheric pressure upon the whole surface of the body. This pressure is so much the more strong, relatively speaking, as the rarefaction of the expiratory medium is more considerable. Under the influence of this external pressure the blood is caused to flow concentrically from the periphery of the body towards those regions where the pressure is less great, namely, towards the respiratory organs which alone communicate with the rarefied air. In consequence of this, while the artificial medium acts by attraction upon the air and pulmonary vessels, the normal air in which the person is placed acts by pressure upon the circulation of the periphery, these two conditions combining to overfill and dilate the blood-vessels of the bronchial tubes and lungs. Hence the possibility of vascular rupture and hæmorrhage in those persons who have a special liability to this complication on account of abnormal weakness of the blood-vessels. In short, what creates the danger in these circumstances is not so much the diminished pressure in the expiratory medium considered alone, as the difference which exists between the pressure at the periphery which is normal, and that within the chest which is diminished.

It should be observed, by anticipation, that this difference does not exist when the person lives in a rarefied atmosphere, as in high climates. The pressure at the periphery being then diminished to the same extent as that within the chest, the relation between the two is not altered. Instead of a flow of blood taking place from the periphery to the centre, there is a general expansion of the blood-vessels at the periphery, and hence the pulmonary vessels are not more full of blood, nor is the danger of hæmorrhage greater. Thus two physical conditions which seem at first sight equivalent have effects radically opposed to each other. Expiration in the rarefied atmosphere of machines has undoubtedly a tendency to produce hæmorrhage, while living in the rarefied air of elevated places is completely harmless in this respect. It is many years since I established the truth of this fact, and that such a condition is free from danger, even though the rarefaction of the air may far exceed that of the air in the machines used.

The predisposition to hæmoptysis, whether shown to exist or only in probability doing so, is thus an absolute counter-indication to the employment of machines with double action, producing inspiration of compressed and expiration of rarefied air. Such an exception, however, being very properly made, I do not hesitate to say that this method of treatment is the best which can be chosen, since it fulfils more completely than the other the indication given by respiratory inaction. In its mechanical effects it is superior to the method in which compressed air is used, while it has the same beneficial effect upon general nutrition, though the latter effect, it need scarcely be said, is less decided in transportable than in fixed machines. There is, therefore, no doubt as to the relative value of these two methods in the conditions which we are now considering. The machines with double action, although of a certain size, are still transportable, so that their employment is not irreconcilable with a country life; machines of this kind are numerous, and those of Waldenburg and Schnitzler should be specially mentioned.

Whichever method be employed, an advance should be made by gradual changes in the use of transportable machines. Beginning with one sitting on each day, having an average duration of ten minutes, at the end of some days, should this practice cause no fatigue, the duration might be progressively extended to a time twice as long. Still later, if it be thought necessary, this should be done twice on each day, but the effects of the treatment should always be carefully watched, and the number and length of the sittings should be diminished if there exist in the interval any sensation of respiratory fatigue or thoracic distension.

I strongly recommend the employment of aërotherapeutic treatment; its good effects quite make up for the slight inconvenience attending its daily use, since there is nothing which equals it in developing the functional activity of the lungs. One circumstance alone, according to my experience, renders this treatment useless and makes it unnecessary; I mean residence in an elevated region, of which the altitude exceeds 4000 feet (1200 *mètres*). Except in this condition, and in those persons in whom the prophylactic treatment of phthisis seems unnecessary, the question of aërotherapeutic treatment should be taken into consideration, and if its employment seems desirable, there should be no delay in putting it into practice. A fault which may have serious consequences is committed by the physician who neglects to employ this method when it seems likely to be beneficial, since one of the most powerful arms in prophylactic treatment is then left unused.

I have only as yet spoken of the hygienic and aërotherapeutic means which may be employed in the treatment of this disease. This is by far its most important part, being the least known, or at any rate the least employed; while, without wishing to detract in any way from the value of pharmaceutic agents, I do not hesitate to assert that it is more powerful in its effects and most productive of favourable results. Even more than this may be said. The hygienic treatment, of which I have been so careful to give all the necessary details, is often of itself sufficient to render the prophylactic treatment effective, so long as it is completely put into practice at the necessary time. Thus during this premonitory period medicines taken by the mouth are only advantageous and opportune in the two following circumstances: firstly, when the hygienic plan of treatment is deprived of its fundamental means of action, namely, permanent residence in the country; in this case the anæmia determined by a town life, and which will become fatally aggravated in persons who are in other ways predisposed to tuberculosis by their constitutional condition, must be opposed; secondly, when the preventive obligation which results from the predisposition to phthisis is associated with some indication furnished by a disease actually in evolution. Two pathological conditions come within the scope of this remark, being associated with the prophylactic period of tuberculosis by a connection which observation shows to be most frequent, namely, pre-existing anæmia and scrofula. It need scarcely be said that these two conditions may be combined, and the indication drawn from their existence be presented by those patients who continue to lead a town life. In this case the definite effects of prophylactic treatment are seriously compromised, and medical treatment, properly so called, becomes of the first importance.

To sum up, during the period which we are now considering, hygienic management, if fully carried out, specially as regards residence in a mountainous region, alone suffices in many cases to give the effect of prophylactic treatment. When this is not the case, certain forms of medicine may be used in its place, and on account of the anæmia or scrofula which exists, cinchona, iron, arsenic, or cod-liver oil might be employed. Cinchona would be generally given in the form of wine, taken either during the course of meals or at their conclusion, and not when the person is fasting, so as to prevent the gastralgia which the continued use of this remedy is likely to provoke when it is thus administered. Neither do I recommend that it should always be given in the same medicinal form. The wine should at certain intervals be replaced by the extract, either in the form of the soft extract administered in pills, or of the dry extract or salt of Lagarraye, which may be employed in lozenges (*cachets*). By means of these simple precautions the bark may be employed as long as is thought advisable, without any reason for its abandonment owing to the distaste of the patient. In the very young, syrup of cinchona should be substituted for the other preparations.

On account of the constitutional debility which is the origin and substratum of every form of phthisis, it might be supposed that preparations of iron are invariably indicated. This, however, is by no means the case, either in the confirmed disease or during the prophylactic period. The debility which leads to phthisis results from impairment of general nutrition, of which the immediate cause is unknown; and this malnutrition, as I have remarked elsewhere, may be perfectly independent of anæmia, in which the number of blood globules is diminished. Since it is only in globular anæmia that treatment by ferruginous preparations is rationally indicated, their employment should be limited to such cases, which indeed are by no means uncommon. I myself usually then administer the iodide of iron, either in the form of syrup for young children, or of pills in other circumstances. These are taken with meals, varying in number according to the individual affected, from four to ten daily. The reason for preferring this preparation is the fact that it contains iodine, which cannot be altogether without effect, though at times it has to be relinquished on account of its astringent effect. There is no need in this case to abandon ferruginous preparations, the tartrate of iron and potash being then most suitable, which, as I have often observed, produces this inconvenience to a much less degree.

Since this salt can be easily thus administered, I prescribe pills of which each contains a grain and a half (10 *centigrammes*) of the tartrate, giving them in the same number as those containing the iodide. Abandoning the wine containing cinchona, I add to the iron salt an equal quantity of the soft extract of that bark, and in this way the patient takes daily from six to fifteen grains (*de quarante centigrammes à un gramme*).

When some improvement shows itself in the anæmic condition, or the special cause of the complaint no longer exists, even though disordered nutrition may still indicate the suitability of treatment which promotes its activity, I begin to administer arsenic in the form of granules containing gr. $\frac{1}{70}$ (*un milligramme*) of arsenious acid. These are taken at the commencement of the two principal meals, two being taken at first on each day, while the daily dose is increased by two granules in each week until it gradually reaches to six, eight, or ten granules, the increase to be at once discontinued should the slightest sign of intolerance present itself. I continue to give arsenic in the largest dose borne by the patient until some improvement of nutrition shows itself, as proved by notable increase in the weight of the body, two or three months being usually sufficient to effect this. I then diminish the dose week by week, and finally cease to prescribe it. After an interval of from six weeks to two months arsenious acid may be again given, this being managed with the same caution as before. The effects and symptoms of intolerance will be discussed at greater length in studying the treatment of confirmed phthisis, in which it also occupies an important place. I will only add here that arsenious acid is beneficial not only as regards the fundamental indication drawn from existing malnutrition, but also in counteracting the nervo-vascular excitability which so often exists in what has been termed the prophylactic period of the disease.

Cod-liver oil forms an indispensable part of the preventive treatment in those patients who are affected by scrofula. The most certain means of protecting these from tuberculosis is by curing the scrofulous disease, and cod-liver oil, administered in sufficiently large doses and perseveringly, is undoubtedly the most effective remedy. It would be useful to associate with it cinchona bark or iron, when the existence of globular anæmia (*l'anémie globulaire*) shows that these would be suitable. In that case, if the patient does not occupy a mountain residence, it would be well for him to utilize the fine season of the year by living at the seaside, or to employ hydrotherapeutics combined with the use of sea-water, according to the indications of the particular case.

Even in the total absence of scrofulous manifestations, I recommend the use of cod-liver oil in those persons who have suffered from whooping-cough, measles, or any other complaint which is liable to produce glandular engorgement, either in connection with the viscera or skin. I have mentioned the special danger to which this would expose the person, and from which it is important to free the organism; the remedies which I recommend, together with the hygienic means of treatment, furnish the best means of attaining this end. When the affected foci can be reached by counter-irritant applications, tincture of iodine and repeated blisters should be perseveringly applied; nothing, in short, should be neglected which could in any way help to produce the desired effect. In case of failure or of cod-liver oil not being tolerated, preparations containing iodine or iodides should certainly be used, which often have a powerful effect in causing the resolution of glandular enlargements, whether inflammatory or scrofulous. Baths containing sulphurous acid are often a useful auxiliary in the treatment.

There may often be occasion to complete the prophylactic treatment by the use of mineral waters, anæmia and scrofula being the two complaints in which these would be specially suitable. This fact need only be mentioned at the present time, and will be reconsidered when the question is discussed *in toto*, as to the employment of mineral waters in the treatment of phthisis.

The hygienic measures which form the essential portion of prophylactic treatment are the best preservatives from a chill, and consequently from the laryngo-bronchial inflammations which are its consequence. At the same time this preservation is not absolute, and is totally ineffective when the hygienic measures specially indicated are but incompletely carried out. In this case the occurrence of such complications in the course of the prophylactic period is to be expected. Such incidents call for constant, or it may be said exaggerated, care even in the slightest forms of the disease. Of no importance, and perhaps requiring no treatment in a healthy person, in one predisposed to tuberculosis they may become chronic, continuing for any length of time, and determining both the development and position of the first manifestations of the disease. Consequently none of these intervening complications should be neglected, however insignificant they may appear. They should be perseveringly opposed until the last sign of their presence has disappeared, and to attain this end there should be no hesitation in employing blisters, friction with croton oil, thapsia, or other counter-irritants, though in laryngo-bronchitis of the same intensity developed in other conditions such means might be totally unsuitable. Such vigilance and anxiety are even more indispensable when the complication is some form of pneumonia. It is then specially necessary to contend unceasingly until resolution occurs, since the persistence of some remnant of pneumonia is the most grievous consequence that can ensue in those whose constitution predisposes them to phthisis.

With these precepts I conclude the study of prophylactic treatment. It is at this time that medical care is powerful and effective. I strongly enjoin its employment, and that unceasing efforts be made in every case to put into practice the different measures which I have advocated. Long experience enables me to be assured of their salutary effects, and the importance of the result well repays, in my opinion, the trouble which must be taken to put the different means which should be used into actual practice.

CHAPTER VI.

TREATMENT OF THE ORDINARY FORM OF PHTHISIS.

Relations between the apyretic phase of its commencement and the prophylactic period.
Fundamental sources of the indications which it presents: constitutional malnutrition; inaction of the lungs; local affection; fever.
Indication drawn from constitutional malnutrition and inaction of the lungs.
Hygienic rules—Hydrotherapeutics—Aërotherapeutics—Exercise of the lungs.
Diet—Milk—Koumiss: its composition and effects—Mode of employment.
Medicinal treatment—Indications and counter-indications of iron—Cod-liver oil—Necessity and good effects of large doses—Means of administration—Glycerine and its effects—Alternation of these two medicines—Arsenic: its effects—Signs of intolerance—*Resumé*—Scheme for daily treatment.

THE treatment of phthisis in its ordinary form will now be considered. This disease is characterized by the slow development of tubercles in a granular form, and has a duration of many years. This is the typical, and at the same time the most frequent form of pulmonary phthisis.

The outbreak of this complaint is characterized by the appearance and persistence of abnormal stethoscopic signs at the apex of one or both lungs. These signs indicate either a permanent catarrhal condition in the upper region of the lungs, a relative absence of permeability with induration of the same parts, or dry pleurisy limited in extent which precedes other symptoms of disease. So long as no physical signs existed, the person, whatever his state of health might otherwise be, was only threatened by tuberculosis. From the moment that this happens he is definitely affected. However slight such manifestation may be, it denotes the realization of the dreaded evil, and presumption has now given place to certainty. This transition is undoubtedly serious, but should not produce discouragement, and still less therapeutic inaction. It simply implies the necessity of more active and persevering treatment; for at this moment, as in the subsequent course of the disease, medical treatment is by no means useless, but still possesses powerful resources.

At the outbreak of the disease this change usually causes no alteration in the general condition of the patient. It produces no new symptom except a more or less frequent dry cough, and the complaint remains free from fever for a variable length of time. So long as no fever exists, the condition of the patient as regards treatment is really the same as during the prophylactic period. More active remedies may be brought into use, since an existing disease has now to be treated, and not a possible evil to be prevented. The fundamental means, however, of accomplishing these objects are the same, since the principal indications are identical. The following assertions which I have made as to the origin and progress of tuberculosis should be remembered, viz. that caseation is a consequence of debility; that the formation of tubercle is a consequence of debility; that accidental irritation of the respiratory organs causes aggravation of the tubercular lesions, and that fever is a process of consumption. These principles point out clearly the sources of the primary and predominant therapeutic indications, namely, the weakness or constitutional malnutrition, the local lesions and the fever. Consequently, so long as pyrexia is absent, the organic debility and local lesion need alone be considered, and the condition is the same as before with respect to preventive treatment.

When pyrexia occurs the conditions change. Not only is the state of the patient seriously modified thereby, but a fresh indication arises which will persist as long as this symptom continues. A new stage now exists in the treatment, new necessities are imposed upon it, and this element furnishes, in my opinion, the most useful mode of division in the complaint, and the most fruitful in its therapeutic results, and should, in my opinion, be always taken as our guide in appreciating the prognosis of the disease.

In the primary apyretic period the indication drawn from constitutional debility and insufficient nutrition forms the leading indication exactly as in the so-called prophylactic period, and its importance becomes even greater, if possible, when pulmonary lesions can be certified to exist. What, in fact, is the end to be gained, or at least to be attempted? This question I would answer to-day in the same words as those previously used in my Clinical Lectures: Improvement of the nutrition and strength should be acquired, so that increased power of resistance on the part of the organism may prevent the development of local lesions, and substitute a stationary condition, or even a process of repair for necrobiotic evolution (*l'evolution nécrobiotique*). Thus the state of the patient, so far as his general condition is concerned, is absolutely the same; there is, in fact, no difference except as regards the indication drawn from the local lesions.

It will suffice to say that the means of treatment are identical in the so-called prophylactic period and in the apyretic phases of the confirmed disease.

The hygienic measures which have been so fully considered should again be mentioned as absolutely imperative, and be unreservedly carried out. All of these from the first to the last should be maintained or applied with unswerving rigour. There is no longer time for discussion or delay; what was advisable has now become urgent, what was opportune is now necessary. No modification need, therefore, be made in the precepts already laid down—permanent residence in the country, and if possible in mountain air, constant watchfulness as to ventilation, methodical employment of the day in walking, making ascents of known distance, and bodily exercise, are now specially required. Such hygienic treatment is as necessary, or even more so than before, and the desired improvement can only be effected by adopting these measures.

The hydrotherapeutic measures which I recommended should be employed with equal rigour. If the person has already been subjected to prophylactic treatment, and therefore for some time accustomed to the cold shower-bath, since the primary lesions in the lung advance but slowly and are unaccompanied by pyrexia, as is usually the case in ordinary phthisis, no modification is required in the plan of treatment hitherto adopted. Its beneficial effects are still maintained as regards the system being fortified against the effects of cold, the activity of the functions of the skin, the exercise of the lungs, and the general stimulation of the organism; and experience enables me to affirm that when this is done the frequency of the cough, so far from being increased, is often notably diminished. This result is due to the effect which the expansion of the peripheric circulation produced by reaction has upon the circulation in the deeper parts. If, on the contrary, the patient has received no preventive treatment, and is seen for the first time in the above-mentioned conditions, that is to say with circumscribed pulmonary lesions, of recent date and unaccompanied by fever, the advantage of hydrotherapeutic treatment is still certain; but since the person is not accustomed to the practice, it should be but gradually brought into use, as previously explained, by means of warm lotions, succeeded by tepid shower-baths to be gradually cooled. If the rules which I have indicated with respect to dry friction be followed, and the change in the lotions and shower-bath be gradually effected, the cold bath can almost always be brought into use after a certain length of time. Insurmountable difficulties will only be encountered in persons who are specially delicate and excitable, and it will then be more prudent to restrict such treatment to the application of cold or perhaps tepid lotions containing aromatic vinegar or alcohol. This counter-indication has been already mentioned while the prophylactic treatment was being discussed.

When once the cold shower-bath has been employed, it may be continued or given up without fear of mistake, according to the mode in which the reaction occurs. This should be instantaneous, strong, and persisting. As long as this triple quality is presented, the shower-bath should be continued perseveringly, it being certain that the treatment is well borne, that the patient is strong enough to endure it, and that such treatment is itself beneficial. If, on the other hand, the reaction becomes less decided both in intensity and duration, and observations repeated during a period of some days show that this modification is persistent and not accidental or temporary, the treatment should be changed, the shower-bath abandoned, and the milder plan adopted of applying lotions by means of a sponge. The occurrence of fever should also cause the cold shower-bath to be abandoned, but it will be seen that this symptom in no way counter-indicates the other modes of applying hydrotherapeutic treatment.

When the patient does not live at a high altitude, aërotherapeutic treatment, either by means of the pneumatic cabinet or a transportable machine, should be practised during the whole continuance of the primary apyretic period. If no counter-indication exists, owing to a tendency to hæmoptysis, preference should be given to transportable machines having a double action, and by means of which inspiration of compressed and expiration of rarefied air can take place. The advantages of this method have been already discussed, and need not be again mentioned. As regards actual results, I have made the following observations: The practice is never without effect, some more or less decided result being always obtained. The least of these effects is considerable delay in the extension of the lesions. These remain limited to their first size for a much longer time than usual, that is for months or perhaps years, and the difference is always so accentuated that it may positively be attributed to the effect of the treatment, since its slowness of extension is far preferable to the oscillations presented by similar lesions when left untreated. In addition to this effect, which I believe to be constant, at the end of a period varying from six weeks to three months a real diminution in the extent of the pre-existing changes is sometimes observed, the peripheric parts of the affected region gradually return to their normal condition, and the lesion is reduced in extent to a central zone, which may barely equal the half of its primary size.

Lastly, in some even more favourable cases, which, however, must be looked upon as exceptional, the abnormal stethoscopic signs are found to totally disappear, and the upper regions of the lungs to be restored to a perfectly healthy condition. It seems to me that this fortunate result cannot be expected to occur in those forms of the disease which are characterized from the first by the existence of solid induration. In the two cases which I have observed, the condition was one which might be more rightly designated catarrh of the apices, and it is also to this variety that the examples of cure reported by Sommerbrodt, Cube, Schnitzler, and others may be referred. At the same time, this distinction must not be looked upon as absolute, since Lazarus has obtained equally favourable results by means of the pneumatic cabinet in chronic pneumonia of the apex, that is to say in forms of the disease which are characterized anatomically by more or less solid induration.

In short, continued maintenance of the lesions in a stationary condition, more or less marked diminution in their extent, or even their complete disappearance, may be expected if the practice of aërotherapeutics be properly carried out. The first of these results seems to be constant, the second frequent, the third wholly exceptional, such effects being obtained more rapidly and with greater certainty in catarrh of the apices than in consolidation of the same parts, though the latter condition does not seem a certain cause of failure. In all circumstances, as is natural, the effect of aërotherapeutics is more pronounced in proportion as the lesions are of less size and extended over a smaller region, that is to say, are more confined to the apex of the lung.

When, for any cause, this method of treatment cannot be utilized, some compensation, slight though it must necessarily be, may be found in gymnastic exercises, in ascents methodically made according to the rules already laid down, and above all in the direct exercise of the lungs by means of inspiration and expiration practised at such an altitude as to insure complete action of the muscular power employed in breathing. In most cases this practice should be repeated twice daily. On the other hand, the habit of singing should be entirely abandoned, as well as of playing wind instruments of music, and this proscription is absolute in character. It has been said that when practised to a limited extent, and quite at the commencement of the disease, such a habit furnishes the means of exercising the voice and breathing, and that this has the good effect of expanding the lungs and increasing the amplitude of the thoracic movements. In making this assertion, however, the result of such exercise is misunderstood. It consists really in the repetition of artificial efforts which have the effect of separating the movements of inspiration as much as possible, of prolonging to an exaggerated extent the duration of expiration, and of producing short and superficial inspiratory movements, which are but little favourable to a proper distribution of air throughout the lungs. This in my opinion is but a harmful disorder of the respiratory mechanism, and a cause of fatigue and irritation; besides which, the tendency which it has to produce hæmoptysis is unfortunately a fact but too well established.

During the period which we are now considering, dyspepsia does not usually exist, or if so to a very slight extent. Bitter medicines, absorbing powders, alkaline waters, or perhaps, on the contrary, mineral acids in very small doses, are alone required by way of treatment; such dyspepsia, when it exists, not being incompatible with the ordinary diet. This should, therefore, be the same as during the so-called prophylactic period, artificial and supplementary means of nourishment being reserved until a later period, when dyspepsia becomes the real difficulty in the treatment. Animal food should, therefore, be principally though not solely taken, farinaceous food being allowed in small quantity, or if possible abandoned altogether. Fatty substances should be as much as possible avoided, not only on account of their indigestible properties, but also because of the medicinal treatment which ought now to be employed. The nourishment taken should not be limited to one form of food. On the other hand, care should be taken to combine simplicity in composition with sufficient variety to prevent distaste. Should the patient live in a mountainous district, and adopt the hygienic measures already mentioned, the question of food will present no difficulty from the appetite remaining in a satisfactory condition, and dyspepsia, should it exist, is a far later and less serious event than in other conditions of the disease. This is by no means the least advantage of such a residence as I undoubtedly recommend. A good red wine, free from acidity, should form part of the diet, even in the case of those persons who as yet have not been accustomed to take it. In such cases the habit must be gradually acquired, and the many real or imaginary difficulties which will certainly be made must be gradually overcome. Any plan that seems likely to do this should be adopted, since wine has in itself a tonic and healthful effect, and when taken in small quantity facilitates the digestive functions. If, however, true repugnance and positive intolerance exists with regard to the ordinary red wine, before giving it up, Madeira, Port, Sherry, Malaga, or other forms of wine should be tried, either in a pure or diluted condition according to the case. Should these means also fail, different varieties of ale should be tried, preference being given, if possible, to that which is made from malt; and if this cannot be taken, some drink containing brandy should be administered, the requirement being to introduce a certain quantity of alcohol into the daily food, should this not have been previously done. Fortunately, these difficulties are unusual, and if the patient is already accustomed to wine, it is advisable to add at once to that already taken the wine of Cinchona bark, administered during or after meals, the vehicle chosen (Bordeaux, Malaga, Sherry, etc.) depending on the toleration and taste of the individual concerned.

The meals should be two in number, so that the patient can take daily in the morning and afternoon a certain quantity of milk, which should be increased gradually to nearly a quart (*un litre* = 35 oz.), if neither the appetite nor digestion are affected by it. If possible, the milk should be drunk in the cowshed at the time of being drawn, and it is only when the repugnance to this method cannot be overcome that it should be taken cold. In addition to its favourable action upon the digestive functions and secretion of the kidneys, milk taken in this way has the great advantage of diminishing the frequence and intensity of the cough, and after a time has undoubtedly a sedative effect upon nervous or vascular excitability. Frequently repeated observations enable me to make this statement with certainty, and such effects are more rapidly produced and more decided when the patient is accustomed to take milk in the cowshed itself, and to remain there for some minutes, the cowshed being supposed to be kept irreproachably clean. I cannot explain the special action of such an atmosphere upon the organism, but it undoubtedly has the effect of calming laryngo-bronchial irritation, diminishing the frequency of the cough, and after some days the sedative effect remains during the intervals which exist between the times of inhaling the above-mentioned atmosphere. This effect, then, must be added to that of the milk itself, and explains the legitimate importance attributed by our predecessors to the prolonged sojourn of phthisical patients in sheds occupied by these animals. No excess, however, should be committed in this matter. Remaining for any length of time in such an atmosphere, however good of its kind, is harmful, from the fact alone that it is irreconcilable with the fundamental principle of treatment, namely, that the air breathed should be pure and frequently changed. Breathing this air during a length of time which varies from three-quarters of an hour to an hour daily, and divided into four or five different periods, is perfectly compatible with other therapeutic requirements, and as I am convinced of the advantage of this practice I recommend it without hesitation, however surprising such an assertion may be. My conviction is so strong that I give this advice even to patients who are unwilling or unable to take milk. Such cases undoubtedly exist, though they are uncommon; and in making this recommendation, I at any rate feel sure that, while losing the complex advantage derived from milk, I obtain the calming effect which this atmosphere has upon the cough and bronchial irritation.

Each step made in advance with regard to the treatment of phthisis again proves the necessity of country life, since, in addition to its own healthful effect, such a residence is as it were an indispensable condition for the possible and complete adoption of the most important therapeutic measures.

Milk also offers a useful resource in the case of persons who absolutely refuse to take alcohol in its ordinary forms. It is then sufficient to add a few teaspoonfuls (*cuillerées à café* = m. 80) of some alcoholic liquid of good quality, either brandy, rum, or kirsch, to the glass of milk taken in the morning or afternoon. Such a drink is agreeable to the taste, as much alcohol being thus administered as possible—an agent which facilitates the digestion of milk, and diminishes the temporary meteorism which milk is sometimes apt to produce. I should recommend the same being done when such intolerance of the digestion exists that it seems necessary to abandon this valuable remedy. If the above plan is adopted every difficulty often disappears, and even should it be otherwise, the addition of lime-water or a small quantity of a mineral alkaline water should be tried before milk is finally abandoned.

It must, however, be confessed that patients exist who cannot be brought to tolerate milk, while again there are others whose residence does not admit of the milk cure being carried on in the cowshed as already explained, while others again may be so weak that the milk must be mixed with some alcoholic liquid of greater strength than those already named. Kouïniss furnishes the means of fulfilling the different requirements which may exist.

Originally employed in Tartary, where it has been in use (according to Landowski) since the thirteenth century, koumiss is the fermented milk of mares which live in a state of liberty on the steppes of Kirghiz. Many establishments exist in Russia for the preparation of this product, and adoption of the special plan of treatment based on its use. It was long thought that the mares of this region produced a milk having special properties owing to the nature of the country which they inhabited, and that the origin of these animals was a point of fundamental importance. The researches of Biel, however, point to another conclusion. The region inhabited is in itself of little importance, while the life led, combining liberty with absence of work, is the real cause of the milk having a special character. Whenever these conditions are realized, mare's milk, from whatever place it may come, differs from all other forms in the chemical composition of its casein, which bears a striking resemblance to that of human milk. This resemblance ceases to exist if the animals are shut up and obliged to do work. These interesting facts show that in places where cow's milk must be used in the preparation of koumiss, in order to assimilate it as far as possible to the typical product of Tartary, the animals must be kept in a state of liberty, and in conditions resembling their natural state. When these precautions are taken, the koumiss prepared from cow's milk is found to have the same properties as that made from the milk of mares, so that this therapeutic agent can be employed in the most different countries, while but a few years ago it was deemed necessary for the patient to make a journey which was often impossible before this remedy could be employed. From a theoretical point of view, the milk of the ass is superior to that of the cow for the preparation of koumiss, since, in containing but a small quantity of butter and casein, and a large amount of sugar of milk, it closely resembles mare's milk. It would, however, be too great a restriction to limit this form of treatment to the employment of koumiss made from ass's milk; and such a restriction is unnecessary, since by modifying the means of preparation it is easy to increase the proportion of sugar of milk in the koumiss of cow's milk, and to separate a great part of the casein which exists in excess by filtration repeated so long as coagulation takes place in the liquid.

It is according to these principles that koumiss is now prepared from cow's milk in all the large towns of Europe, such as Paris, London, Berlin, Vienna, etc., and in the circumstances which I have just mentioned constitutes both for the physician and the patient a resource of great value.

In order to appreciate the advantages derived from this substance, it is sufficient to know the modifications caused by the process of fermentation. The production of alcohol, of carbonic acid in a free condition or dissolved, of lactic acid, and the elimination of a large part of the casein, are the fundamental results of the process. The proportion of alcohol varies from 2 to 3 per cent., having been 2·253 and 3·031 in the koumiss derived from cow's milk and analyzed by Kokosinsky, and 1·7 to 2 in the koumiss of mare's milk analyzed by Stahlberg. The total proportion of carbonic acid varied from ·6 to 1·3 in the koumiss of cow's milk prepared at Paris, which has just been mentioned, and from ·9 to 1·1 in the product examined by Stahlberg at St. Petersburg. This observer has also shown that the quantity of these elements varies during a certain number of days after the time of its preparation, the amount of alcohol increasing daily from the first to the sixteenth day, while the carbonic acid gas in solution does not vary in amount to any notable extent after the second day, though the quantity of free gas exists in largest quantity towards the fourth day, having been found at that time to the amount of ·966 per cent., whilst on the sixteenth day it was only ·799. The lactic acid varied between ·702 and ·887 per cent. in the koumiss of Paris, whilst in that of St. Petersburg the amount increased from ·475 per cent. on the first day to ·831, the quantity which was found to be present on the sixteenth day.

On account of its special composition, koumiss has an acidulous and alcoholic taste, which in no way recalls the taste of milk. When taken, it does not leave the clammy and bitter sensation in the mouth which renders milk disagreeable to so many persons, and I have not really met with a single patient whose repugnance to it lasted more than two or three days. Usually it is taken readily from the first.

Digested without difficulty, koumiss has not the same inconvenience as milk with regard to meteorism and flatulence.

It maintains and excites the functional activity of the digestion when administered in the place of milk at the commencement of the disease in such small quantity as two or three glasses daily. It also prevents and counteracts the effects of dyspepsia by means of the alcohol and carbonic acid which it contains, and has a stimulating effect upon the whole organism, without producing excitement; and when it is taken in large doses or forms the whole diet, as often occurs in the advanced stages of phthisis, shows its effect upon the nutrition of the patient by increasing the weight of the body, and its influence over the processes of organic combustion by the modifications which it produces in the secretion of the kidney. The researches of Biel, in fact, have shown that when nourishment is carried on solely by means of this product, the daily excretion of urea, phosphoric and sulphuric acid is increased as compared with the quantity excreted anteriorly or subsequently, while on the other hand that of uric acid is notably and rapidly diminished. At the same time a reduction of temperature occurs, and improvement takes place in the local and general symptoms.

It is specially when tuberculosis, having reached the period of softening and pyrexia, is complicated by dyspepsia, which interferes both with nutrition and the administration of medicine, that the treatment by koumiss alone may be of important service, at any rate when the pyrexia does not persist. Five or six weeks will then suffice that this complication may be cured, the quantity administered varying according to the case. After this time, as a rule, the dyspepsia will be relieved, the fever have diminished or ceased, the patient will have gained both strength and weight, and the local lesion have improved owing to cessation of catarrhal or inflammatory complications. The benefit is, therefore, twofold: the lesion itself has improved, and nutrition is again possible, as well as the employment of whatever medicinal treatment is thought advisable. In the above-mentioned conditions, nothing, in my opinion, is more advantageous than the exclusive administration of koumiss continued for some weeks. On the other hand, when the disease, though having reached the stage of softening and pyrexia, is accompanied by persistent dyspepsia, I should not recommend this uniform plan of treatment. So long as it is tolerated, the food habitually taken is, in my opinion, more beneficial. Medicinally, other substances appear more advantageous, and these I should continue to give though enjoining a mixed form of diet, in which koumiss is taken in the place of milk, which is rarely well borne by those who inhabit towns. This should be taken to the extent of three, four, or five glasses daily. The stimulating and nutritious effect of this substance is now of real benefit. Unlike milk, koumiss should not be administered in the early morning on account of its alcoholic properties, but in the interval between meals, and as long as at least an hour before the time when food is taken.

When the disease at an earlier stage is still in that apyretic period which has been so far specially considered, koumiss should only form part of the treatment, being employed in the same quantity as I have recommended with respect to milk. In my opinion it should be administered: (1) when the repugnance to or intolerance of milk cannot be overcome; (2) when the patient is already suffering from serious debility; (3) and specially when the patients inhabit a large town. In this case there is difficulty in taking milk, according to the rules which I have laid down, which is apt rapidly to produce some disorder, or, at least, a sensation of discomfort in the digestive organs; and since residence in town causes symptoms of dyspepsia to show themselves more early and with greater certainty, there is every advantage in giving koumiss in such cases. Its effects are in many points similar to those of milk, and the stimulating influence which koumiss specially exercises is by no means to be despised with regard to the inhabitants of towns. By its exciting effect upon the digestive functions, it may delay for a long time the development of the more alarming period of dyspepsia. One fact, however, should be remembered, which may be considered certain, that as regards this or other complications nothing is equal to residence in a mountainous district, and complete observance of the hygienic rules already laid down.

This, then, is the hygienic treatment of confirmed but apyretic tuberculosis, and the medicinal treatment which influences the fundamental indication, viz. insufficient nutrition, or constitutional debility, will now be considered. It will be seen that this differs but slightly from the prophylactic treatment already discussed.

The occurrence of the first pulmonary lesions is, in my opinion, no reason for discontinuing the use of iron, at any rate if anæmia exists. There seems but one counter-indication to its use, though this is absolute, namely, the previous occurrence of hæmoptysis, or perhaps that special form of constitution, already mentioned, which must create a fear that this is impending. The patients in whom such is the case, as already said, are impressionable, with a fine semi-transparent skin, and delicate, visible veins. In them the cardio-vascular system is in a persistent state of irritability, which the least influence may exaggerate, and the vessels of the head become suddenly and repeatedly congested. In such cases iron, by increasing this cardiac irritability, is apt to produce congestion of the bronchial tubes and hæmoptysis. Some eminent physicians have absolutely condemned the use of iron in those who either certainly or probably are affected by phthisis, from the idea that it increases the liability to hæmoptysis. This prohibition is, however, too absolute, and the relation of cause to effect which it supposes to exist between the use of iron and hæmoptysis is by no means so definitely established. It seems that the counter-indication already mentioned need alone be considered, and that iron should be administered to all other patients suffering from decided anæmia. To patients affected by scrofula the iodide of iron will be most suitable. In other cases liquid preparations of iron should be preferred: either the ethereal tincture of perchloride of iron, the tincture of Bestucheff in the dose of five to twenty minims in a glass of water sweetened by sugar (*eau sucrée*), the tincture of tartarated iron to the amount of from two to five drachms (10 *à* 20 *grammes*) daily, the solution of perchloride of iron from twenty to thirty drops daily, in two doses, each of which is contained in half a glass of water; or an elixir may be prepared containing brandy or rum, to which some soluble iron salt is added. Of this, the half or whole of a liqueur glass may be taken at the end of each meal, the effects of iron and alcohol being thus combined in one medical preparation.

The treatment by iron should not be continued for more than two months, in which time the troubles due to the anæmic condition are usually diminished. Even should it be otherwise it is inadvisable to prolong its use, but recourse should be had to arsenic, to be administered from the first when there is no reason to give iron; that is to say, in the great majority of cases.

Arsenic, the advantage of which in the prophylactic treatment has already been mentioned, will be soon considered at greater length; but in following the logical sequence dictated by relative importance, the medical agent which seems to be the most powerful in its therapeutic results, namely, cod-liver oil, should be first considered.

It might be thought—and the assertion has been more than once made—that this remedy is only of service in scrofulous phthisis. This, however, is not the case. That it is then more specially appropriate, owing to the ætiological form of the disease, and therefore more rapidly beneficial, is true. This fact, however, does not diminish the advantage of its use in other circumstances. Tuberculosis, whatever the cause of its existence may be, implies organic loss and insufficient nutrition, while cod-liver oil, being specially nutritious, really answers to this indication in every case. In addition to this, on account of its chemical composition it has an alterative, or, as may more truly be said, a resolving, effect upon the local lesions, and even in the least fortunate cases keeps them for a long time in a stationary condition. It is certain that diminution in the extent of the lesions, combined with increase in the weight of the body, may be observed when the treatment can be so completely carried out as to ensure its efficacy. On the other hand, extension of the existing lesions very rarely occurs during the time of its employment. It will be understood that the initial apyretic stage of the disease is alone being considered.

Unfortunately, in practice this medical agent cannot be employed without unceasing difficulties, some existing in reality, and due to the fact that the oil cannot be tolerated; others being imaginary, and caused by repugnance to the odour and taste of the remedy. In addition to this—and attention should be strongly called to the fact, which is but too little recognized—the oil has not the same good effect if given in small quantity. The numerous failures which have caused this plan of treatment to fall into discredit have been mainly due to its having been wrongly employed, namely, in insufficient doses. Though the distaste of the patient may be overcome, though the remedy may be taken with regularity, no advantage will be gained by the daily administration of one or two teaspoonfuls. Both the patient and his digestive powers may be fatigued, without the good effects which larger doses would have produced. The necessity of employing large doses is the real difficulty in the treatment, since on this account the oil must be taken as it is, and certain therapeutic forms, as, for instance, capsules, would not enable it to be taken in sufficient quantity on account of the dyspeptic symptoms which might then immediately ensue.

At the same time, though the adoption of this treatment is difficult, it must not be considered impossible. The beginning is really the awkward time; but when two table-spoonfuls (*deux cuillerées à bouche*) can be taken daily, there will be but little difficulty in increasing the amount to four or even six table-spoonfuls—the least quantity which, in my opinion, is of real service. Even in the impure atmosphere of the hospital a far larger amount may in some cases be taken for months, and this result can naturally be obtained with less difficulty by patients living in the country air; that is to say, in conditions which are most favourable to digestion and assimilation.

In my opinion it is best to proceed gradually, the amount taken at first being one table-spoonful daily: in about a week twice this amount may be administered, and after a second interval the dose may be again increased so as to reach more or less rapidly the amount of four table-spoonfuls. These should be taken in two doses from a small glass, that the patient may have to drink the oil but twice daily. The real difficulty is to take this quantity; but observation has shown that when the dose of four table-spoonfuls is once tolerated, there is rarely any difficulty in increasing it to three or four ounces (100 *grammes*), that is to six or eight table-spoonfuls. Still larger doses may at times be given; and though, on the one hand, some persons have great difficulty in taking the remedy, others have an exceptional power of doing so, and, without exaggeration, can drink cod-liver oil by the glassful, taking as much as from seven to ten ounces (*de* 200 *à* 300 *grammes*) daily.

In 1876, three patients were under my care during many months at the hospital, who had acquired the power of taking without difficulty, and besides their ordinary diet, a considerable quantity of this remedy. For forty-eight consecutive days they took as much as twenty table-spoonfuls (*vingt cinq cuillerées*) or ten ounces (300 *grammes*) daily, and left the hospital in the most satisfactory condition, the local lesions having without doubt diminished in size. Notwithstanding my advice, they have not been again to see me. Such a power of assimilation is certainly exceptional, but these interesting facts show the possibility and advantage of administering large doses, and that no fixed quantity can be considered as the dose. The remedy should, in my opinion, be given in as large a quantity as can be borne, three or four ounces (100 *grammes*) being considered the minimum dose necessary. Improvement in the general condition, diminution of the cough, increase in the weight of the body, which has been observed to be as much as from three to four ounces daily (100 *grammes*), are the signs by which the remedy is known to be doing good, while the absence of indigestion proves that the useful limit has not been passed. In appreciating the value of the remedy, not only must the condition of the appetite and digestion by the stomach be taken into consideration, but also its toleration by the intestine. The occurrence of diarrhœa counter-indicates any increase in the dose given, and, should it persist, threatens a retrograde step in the disease. So long as it exists the nutritious effect of the remedy is absent, and in a few days the progressive increase, which has been observed hitherto in the weight of the body, is replaced by diminution. This fact has been frequently observed, even in patients who have continued to take the oil without difficulty, and in whom the functions of the stomach remain perfectly normal.

The question arises whether cod-liver oil should be taken when the stomach is empty, or combined with meals. In respect to this no absolute rule can be laid down, some patients tolerating it best during the interval between meals, while others only digest it when taken at their very commencement, or during their progress. Such individual distinctions should be respected. The essential point is that it *should* be taken, and all else is but accessory. A uniform rule is, therefore, impossible, and concession must be made to the idiosyncrasy of each individual.

Certain modes of administering cod-liver oil may cause it to be more easily tolerated, and in this respect the following are, in my opinion, the most advantageous plans. If the patient is troubled by nausea, even though he takes the precaution of swallowing food immediately after the oil is administered, he should be recommended to take it while a slice of lemon or orange is in the mouth. If this very simple process is without effect, he should be advised to try a plan which has the advantage of facilitating the digestion of the oil, and of administering alcohol. When the remedy is taken, brandy, rum, kirsch, or whisky should be added to it in the proportion of a third, or a quarter of the amount taken, according to the patient treated. Such a mixture, which only leaves in the mouth a taste of the alcoholic preparation used, is less fatiguing to the stomach than the oil alone. When, as is very rarely the case, this mode of administration fails, the plan of Williams may be adopted, viz. that of adding gr. $\frac{1}{100}$ (1 *milligramme*) of strychnia to each dose of the oil. This combination has no effect upon repugnance to the oil, but much facilitates its toleration by the stomach. With the same object, and in imitation of Foster, who suggested this plan in 1868, a small quantity of ether may be added to the oil; this increases the pancreatic secretion and undoubtedly promotes the digestion of the remedy. At the same time, if care is taken to convince the patient that this plan of treatment is considered of the highest importance; if the changes made are but gradual, which is indispensable; if the idiosyncrasy of the patient as to the time at which the remedy should be taken is respected, it will seldom be necessary to have recourse to these auxiliary measures, since repugnance to the remedy is but rarely insuperable. The real difficulty is far more often due to some disorder in the functions of the stomach, which may be effectually removed by means of alcohol, strychnia, or ether. If, however, for some reason cod-liver oil either cannot be taken or is not tolerated, hydrocarbons must not on that account be abandoned, but glycerine should be employed, the administration of which is not found to offer the least difficulty. This subject will soon be again considered.

When the administration of oil is possible, this remedy should be perseveringly continued, unless some gastro-intestinal complication occurs, until the onset of pyrexia. This symptom is, in my opinion, an absolute counter-indication on account of the change which it produces in the secretions of the stomach. It is almost certain that the oil will be no longer tolerated, but must be now replaced by glycerine, which, on account of its alcoholic character, is well digested even when pyrexia exists. This is an important fact, which I have frequently proved to be the case, even at the hospital, and which a large experience enables me to affirm as true. Another counter-indication is often said to exist, by routine as it would seem, to the employment of the oil, namely, the hot season of the year. This distinction, if it exists, should not, in my opinion, be considered absolute. If, when this season has commenced, the remedy is still found to be advantageous, and the digestion is carried on satisfactorily, the heat is of little consequence, and it should certainly be still administered. The truth is that fat is tolerated less well in summer than in winter, but it is best to wait until the impatience of the organism shows itself by positive signs before giving up one of the most powerful means of treatment. In this case, again, general rules have been laid down, which are but theoretical presumptions, and which must give way to individual observations. If, indeed, the hot season of the year is really found to have this effect, some change should be made; but the time need not be useless to the patient, since glycerine may be given in the place of cod-liver oil, which will be always beneficial and well tolerated.

The cod-liver oil should, in short, be administered for as long a time as possible, and when for any reason it cannot be borne, glycerine may take its place, one or other of these remedies being invariably administered. This alternation is, in my opinion, a real proof of progress in therapeutics. Until the time when observation showed me the advantage of this method, and the good effects of glycerine, my patients were necessarily deprived of the benefits produced by hydrocarbons as soon as the cod-liver oil was no longer tolerated. An experience of ten years has now proved to me that glycerine can replace cod-liver oil with benefit, and that the permanent administration of one or other of these two agents assures to the patient the advantages of a remedy which is specially nutritious in character. When it can be administered, cod-liver oil should be preferred, since, besides its properties as a fatty substance, it has, on account of the iodine, bromine, sulphur, and phosphorus which it contains, a special action which cannot be replaced.

Glycerine, a triatomic alcohol of the formula $C_3H_8O_3$, like other forms of alcohol, is an inhibiting agent, which, however, has also a direct and positive influence upon the general nutrition of the body. This effect may simply result from its increasing the appetite and stimulating the functions of the stomach, or perhaps from its increasing the activity of organic combustion, as the interesting researches of Catillon tend to prove, by showing that under the influence of this remedy a relative and absolute increase of carbonic acid is present in the expired air. This point is doubtful, and in my opinion all these elements should be rather looked upon as joining to produce the final effect, namely, increased nutrition, which has been repeatedly proved to occur by increase in the weight of the body.

I have frequently made the following experiment. The food taken by the patient and his conditions of life remaining the same, I have administered glycerine, according to the process which will soon be described, for a period of fifteen days. At the end of that time an increase of weight, small, it may be, but real, is found to have occurred. I then cease to prescribe glycerine during the same period of time, after which this increase is not found to have continued, the weight being seldom equal to that which existed when the previous measurement was made. Glycerine is then given in the same dose as before, and in a fortnight the patient is found to have regained the weight which he had lost. The other circumstances which might have some effect upon the nutrition of the patient had remained identically the same throughout the observation, the benefit gained being, therefore, undoubtedly due to the glycerine, which substance must possess, in the same way as oil derived from fish, the property of increasing the nutrition and weight of the body, to the advantage of the organism. This fundamental effect, which I term eutrophic (*eutrophique*), is, therefore, common to the two remedies, and justifies the statement recently made, that glycerine can supply the place of cod-liver oil to an extent which is really beneficial.

The glycerine, taken internally, must be pure. Perfectly neutral when tested by reagents, it should be without colour or odour, of sweet taste, and without any after-taste. The daily dose should not approach in quantity that of cod-liver oil on account of its alcoholic effects, which may amount to intoxication. I have observed this in two patients, who, finding themselves in better health on account of the glycerine, took a larger quantity than had been prescribed. The dose I order is from three to four table-spoonfuls (*de 40 à 60 grammes*) daily, the larger dose being reserved for those who present no sign of nervo-vascular excitability. Agitation, unaccustomed loquacity, persistent insomnia, are symptoms which show that the useful dose has been exceeded, while a still more certain proof that such is the case is elevation in the temperature of the body. This is the rule when glycerine is taken by a patient for several days at the dose already mentioned, but the elevation does not exceed from two to four tenths of a degree (F.) as compared with the mean temperature of the preceding period. Within these limits, and even as far as six-tenths of a degree, the difference in temperature simply indicates the physiological action of the remedy, and may be neglected.

When, on the contrary, without the occurrence of anything which might produce pyrexia, the elevation of temperature is found to remain as much as a degree higher than before, and specially if this elevation is still more pronounced when a second change is made in the remedy, this may be looked upon as proving that it is administered in excess. In the adult I have never met with this counter-indication, even when a dose of three table-spoonfuls daily has been given for as many months without interruption; I have, however, done so at times when five, and almost always when six table-spoonfuls have been given, either during or shortly after the first month of its use.

Glycerine may be given alone, but I prefer adding to the daily dose one drop of essence of mint, and two drachms of brandy or rum. I can assert, from having frequently myself tasted it, that this mixture is agreeable to the taste, easily digested, and that, after being taken uninterruptedly for many months, it produces neither a feeling of satiety nor distaste. It may be taken in two or three doses daily, either with meals, or in the interval between them. It need scarcely be observed that the effect of adding this small quantity of brandy or rum is merely to modify the insipidness of the glycerine, and to promote its digestion. The object is not to give alcohol as a remedy, since in the first place the amount given would be quite insufficient, while in the second such an addition would be quite illogical, glycerine being itself an alcohol which adds a nutritious effect to the ordinary action of alcoholic preparations. Such a combination of effects in one substance is the great advantage of this remedy, which, in my opinion, should be administered whenever, for any reason, cod-liver oil must be abandoned. This is my practice, patients under my care invariably taking one or other of these two remedies.

Except in those cases which indicate the requirement of iron, I give arsenic with the same pertinacity during the apyretic period of ordinary phthisis. It is a mistake, in my opinion, to reserve this remedy for those forms of the disease which are associated with arthritis or herpes. Undoubtedly there is then a special ætiological indication which itself suggests the administration of arsenic and ensures its utility, but even when this complication is absent, constitutional malnutrition, the constant and ordinary accompaniment of phthisis, whatever be its origin, indicates the same requirement. The pertinacity with which this pathological element presents itself has brought me to be equally constant in employing a remedy which is capable of combating therewith. The uniformity of my practice is based upon the same foundations, and therefore, in my opinion, arsenic should be administered for the same reason as cod-liver oil and glycerine. The sole question to be solved is whether arsenic really answers to the indication presented by insufficient nutrition. The answer is obvious. The nutritious effect of this remedy is shown by increased appetite, improved digestion, and greater activity in the nutritive processes of life, effects which lead after a time, whether in health or disease, to increase in the weight of the body. It would, therefore, be difficult to find an agent more capable of resisting the consumptive tendency which exists in every form of the disease. Nor is this all, since arsenic also answers to two important symptomatic indications, calming both nervous irritability and cardio-vascular excitement. The latter fact, long established by clinical observation, has been experimentally proved by Unterberger, who showed that when arsenious acid was administered to animals there was in all cases diminution in the blood pressure, and retardation of the pulse. For these reasons I continue to regard the requirement of this remedy as constant in the ordinary form of phthisis, this being already stated in published works, and a more varied and larger experience confirms such a conclusion, which is the rule of my practice.

The association of arsenic with cod-liver oil or glycerine, combined with the mode of life already explained, is certainly the most sure means of obtaining that organic repair which should be the first object of the physician. At the commencement, and in the first stage of tuberculosis, such treatment is not only sure, but infallible in this sense, that after a time, which varies from four to six weeks, a positive improvement is always found to occur in the general condition, the appearance, the strength, and weight of the patient. When this complex mode of treatment is continued for several months there is often observed, specially when the local means, soon to be considered, are at the same time employed, a diminution in the extent of the pulmonary lesions, and a retrocession of the disease. Even when this desirable result is not obtained, as unfortunately but too often happens in towns, the constitutional improvement produced is still beneficial, since it prevents the development of fresh lesions, and permits the patient to live with those which already exist, conferring thus the advantage of a relative cure. Whether the advantage is temporary or definite depends upon the conditions in which the individual lives.

The same plan of treatment succeeds even at the hospital, and no better proof than this can exist of its efficacy. I have frequently had the satisfaction of seeing patients, who were under my care for three or four months, leave the hospital in such an improved condition that their occupation and ordinary mode of life could be resumed without fatigue. The lesions were the same as when they were admitted, but the organism was now able to resist both their tendency to extend and wasting effect, and as much as a year, eighteen months, or two years passed before the patients were again obliged to give up work and seek medical advice. In two cases this interval, which measures the length of the relative cure, lasted as long as three years, and only ended then on account of accidental illness produced by a chill. Such being the possible result, it may easily be conceived that treatment which can effect such a relative improvement, even at the hospital, may produce a definite cure when carried on in a better atmosphere, and in patients who live in such hygienic conditions as are able to increase or prolong its effects. It is then that the improvement in the general condition may be accompanied by diminution or even disappearance of the local lesions, as shown by the more and more circumscribed extent and slighter intensity of the abnormal stethoscopic phenomena. I have seen many examples of this fortunate result when hygienic, medical, and climatic treatment had been continued for two or three years. The last in date was that of a young man, M. C——, who came from the south of Russia, and in whom I found last May that the physical signs which I discovered as usual six months before had without doubt almost completely disappeared.

I have now for some time abandoned the liquid preparations of arsenic, not only from their being inconvenient in use, but from their leading, in my opinion, more than other remedies to such complications as gastralgia and intestinal troubles. Hence I employ almost exclusively granules containing gr. $\frac{1}{65}$ (*un milligramme*) of arsenious acid. It is an essential point that these granules should come from a chemist of irreproachable credit, since their fabrication is such a delicate matter, that unless the necessary time and care are employed in making them, the amount of arsenic contained is apt to vary, so that while some granules are inert, others contain as much as twice the proper quantity. They should be administered at the commencement of the two principal meals; two granules being taken at first daily, while in each week two are added, until as many as six, eight, or even ten are taken, according to the case treated. This amount when reached should be continued for an indefinite time, unless fresh pathological trouble occurs, or some indication exists that the medicine is no longer tolerated—as shown by cramps of the stomach, ophthalmia, cutaneous eruptions, vomiting, or diarrhœa. Besides these well-known signs another may exist, which should never be forgotten, namely, a feeling of fatigue after walking, followed in a short time, if the arsenic be continued, by decided weakness of the lower extremities. This indication is not unfrequently found to be first in date of those which reveal that the system is saturated with arsenic. When any of these symptoms occurs, the arsenic need not be altogether discontinued, but the dose should be sensibly diminished, and only increased at a later date by gradual and cautious changes to the maximum dose which can be tolerated.

It should be finally remarked that this plan of treatment, though apparently so complicated, can be put into practice without difficulty or inconvenience. Thus, if the milk be also taken into consideration, the plan for the day will be as follows: in the morning, and in the afternoon at least an hour and a half before the evening meal, milk should be taken, and at the beginning or during the progress of the meals cod-liver oil or glycerine, and the granules containing arsenious acid. This is the complete plan of treatment, and any other indication which shows itself can be easily fulfilled, such methodical treatment preventing neither the walks nor aërotherapeutic exercises of the patient.

Such, then, are the hygienic, mechanical, and medical means of treatment by which, in my opinion, the fundamental indications furnished by the general conditions of the patient, and the inertness of the lungs, are best fulfilled in the initial period of ordinary phthisis. The indications which may be drawn from the local lesions will be considered at the commencement of the following chapter.

CHAPTER VII.

TREATMENT OF THE ORDINARY FORM OF PHTHISIS (*Continued*).

Fundamental indication furnished by the local lesion—Zone surrounding the tubercle—Catarrh of the apex and its relation to tubercle—Importance of local treatment—Different modes of producing counter-irritation.
Phenomena of catarrh and modifications of the medicinal treatment indicated thereby—Creasote and its effects. Modes of administration—Alcohol and its effects—Sclerotic as opposed to caseous degeneration—Some accessory indications—Medicinal forms of milk.
Fundamental indication furnished by pyrexia—Different causes of pyrexia in pulmonary phthisis—Connection between the form of fever and its cause.
Treatment of pyrexia—Indication of quinine; its effects. Mode of administration—Subcutaneous injection of hydric-bromate of quinine—Indication of salicylic acid; its effects. Mode of administration—New mode of administering salicylic acid.

THOUGH improvement in the general condition of those affected by tuberculosis has undoubtedly a beneficial effect upon the pulmonary lesions, it would be, in my opinion, a serious mistake to expect from such indirect influence alone any real diminution of the local affection. The physician can and should do more. Every tuberculous deposit is surrounded by a hyperæmic or inflammatory zone which seems to increase the extent of the specific lesion. Though not yet tuberculous, it will almost undoubtedly become so on account of the diathesis which exists, if left to take its own course. In reality this zone produces tubercle (*c'est vraiment une zone phymatogène**), and not only precedes, but even causes increase in size of the tubercular deposits. It is, therefore, most important

* See note, chap. i. p. 7.

to prevent or oppose such catarrhal or inflammatory congestion as is the primary effect of tuberculosis in the affected region. Whilst the latter preserves this premonitory character, and remains in a state of simple congestion, therapeutic intervention, which has a real effect upon such a condition, should be brought to bear upon this zone. This is the first, and an important indication as regards local treatment.

Whatever is said of advanced lesions applies equally well to the occurrence of primary changes. The mechanism according to which tubercle forms is identically the same, whether its primary origin or secondary extension is considered. The production of tubercle is not an occurrence which is isolated and independent of all other pathological change, but the result of preliminary irritation expressing itself in the form of congestion, catarrh, or inflammation. Instead, however, of being followed by the usual inflammatory products, this irritation, owing to the character of the affected part, and the diathesis which overrules its effects, engenders the specific lesion named tubercle. Thus it is—and the fact should be noted—that the formation of tubercles occurs in a second stage, which has been preceded for a longer or shorter period by one of congestive irritation. It is this hyperæmia which is the first and generative act. Catarrh of the apex is by no means always associated from the first with the existence of tubercles, being but the indication of imminent tuberculosis, and though having on this account a most important signification, it is by no means proved to be always associated from the time of its onset with already existing tuberculous products. Some facts, on the other hand, show clearly the truth of my opinion, that catarrh is merely at first a premonitory indication. Such forms of catarrh, as I have already shown, may be cured in a few weeks by means of aërotherapeutic treatment, and the rapid recovery in these circumstances must be looked upon as proving the non-existence of tubercular deposit. I myself observed one case of this kind, which seems to show this as clearly as could be done by anatomical demonstration.

The patient was a young man aged twenty-five, of good constitution, and whose health I had frequently been able to assure myself was irreproachable. Towards the end of 1877 a feeling of fatigue, some emaciation, and a slight cough were found to exist, and in February and March, 1878, hæmoptysis occurred on three occasions in the absence of any appreciable stethoscopic indication. As the patient refused, notwithstanding my advice, to return to the country, I employed prophylactic treatment, and it was agreed that he should come each week to be examined. No change occurred, as shown by repeated examination, until the beginning of June, when I found undoubted signs of catarrh at the apex of the right lung, namely, the existence of crepitation in the whole supraspinous fossa, while in the preceding week my examination had given, as usual, a negative result. The catarrh, therefore, which was now found to exist was but commencing. Without more delay, I induced the patient to live in a mountainous atmosphere. Three months later I again saw him in Switzerland, and upon making a most careful examination at two separate times, I found no evidence whatever of catarrh, which was the more surprising since its origin had been preceded by the occurrence of hæmoptysis. Upon my seeing him again a year later, the respiratory system was still, as far as could be judged, in a healthy condition.

The significance of this fact is obvious. Owing to a combination of circumstances, unfortunately but too rare, the catarrh of the apex was recognized at the first moment of its existence. Besides other means of treatment, the influence of living at a high altitude was brought to bear on the disease and in three months a cure was effected. The affection could not have been due to existing tuberculous deposits, but showed that their formation was imminent, being, as has already been observed, a form of catarrh which leads to the formation of tubercles (*un catarrhe phymatogène*). Since it was treated from its very commencement, a rapid and complete cure was obtained; and success would be more often met with in these cases could the physician treat the disease from its very commencement, that is to say, if he were not discouraged by the mistaken idea that catarrh of the apex must be taken as an indication of confirmed tuberculosis. Though it may be so in some cases, this connection does not always exist, my observations having brought me to this conclusion, which I hope others will share with me.

In the same way that the zone which surrounds existing tuberculous lesions is itself in a state of congestion which threatens and may result in the development of fresh tubercles, so catarrh of the apex, which in the great majority of cases is the initial phenomenon of ordinary phthisis, may be simply a catarrh of this nature (*un simple catarrhe phymatogène*), and unaccompanied by real tubercle. During how long a time this benign character may be preserved it is impossible to say, but that such a stage does exist is, in my opinion, certain, and the physician who recognizes the existence of catarrh limited in position to one or both apices should consider and treat it simply as a premonitory indication and curable affection. If the alteration has existed but a short time this is most probably the true state of the case, and even when of older date the treatment adopted may still retard or prevent the formation of tubercle, so that from every point of view too much confidence has a better effect than a feeling of discouragement.

Owing to these facts, the importance attached by me to local treatment can be well understood. This should be employed from the time when the first stethoscopic signs of disease are recognized. Graves, who should be quoted on this, as on many other points, thought that counter-irritation should be permanently practised in all persons in whom the development of tuberculosis is to be feared; that is to say, he looked upon it as one form of prophylactic treatment. I would scarcely say as much as this, though believing that this mode of treatment should be adopted whenever the first signal of a confirmed lesion is recognized, however slight it may be. Without ignoring the great service which such treatment may render in the more advanced periods of the disease, it is, in my opinion, at its first onset that this undoubtedly has its greatest power.

To be beneficial, however, it should be continued with unswerving perseverance. Whichever mode of counter-irritation is employed, one application will be undoubtedly fruitless, the repetition and almost the continuance of its action being the condition of real success. When the truth of this fundamental precept is once recognized, the different modes of practising counter-irritation are almost equally good, and flying blisters, points of red-hot iron, or caustics may be employed according to the case. The effect of blisters upon the mucous membrane of the urinary passages should not, however, be forgotten; and though this fact may be neglected when but one or two blisters are applied, at times separated by an interval of some days, this danger not existing in these circumstances, blisters should only be applied when cowardice prevents any other means of counter-irritation from being employed. The application of heated iron points can be made with ease and rapidity, and these, causing but little pain, can be used in most cases, though at the same time their action is more superficial and less lasting than that of caustics,* to which I have now been led to give the preference.

---

* It will be remembered that the name of cauterization has been given to the action by means of which living tissues can be disorganized, whether from the effect of heat or chemical agents. When heat is employed, the agent is termed the actual cautery, or simply cautery, a chemical agent being termed a caustic or potential cautery.

The plan which I adopt is to apply Vienna paste (*pâte de Vienne**) in three places beneath the clavicle, on one or both sides according to the seat of the lesions; the parts covered being as small in size as a threepenny piece (*une pièce de vingt centimes*) at most, and widely apart from each other. I do not keep up the irritation, or employ either the pea or epispastic ointment, since the suppuration required to eliminate the sloughing part is, in my opinion, quite sufficient. When, however, the part commences to heal, other substances are employed in the same way, this being done as long as any favourable change is observed. Long experience has confirmed the statements which I formerly made as to the beneficial effects of this mode of treatment. The extension of the local lesion is prevented either definitely or for a long time; and not only does this happen, but resolution takes place, or at least the progress of the disease is arrested in this sense, that the lesions do not pass beyond the condition of induration and catarrh, and that the ulcerative softening, which is so important a consequence, is delayed for a long period. Besides this, repeated application of the cautery in the form of points has often the effect of relieving the cough, and diminishing the thoracic pain which is so often felt at this period of the disease.

When an insuperable objection is made to the employment of this form of treatment, thapsia † plasters, croton oil, or, the least effectual of the series, tincture of iodine, may be applied to the part. Such applications, though undoubtedly less effective than the others, are at the same time preferable to inaction. In every condition whatever is possible should be done, and the counter-irritant plan of treatment should never be abandoned, being one of the most powerful arms of therapeutic medicine.

The first symptoms of laryngeal irritation should be treated

* The *pâte de Vienne* is made of caustic potash (five parts) and quick-lime (six parts), made into a paste by being moistened with water.

† Some plants of this genus of Umbelliferæ (*Th. villosa, Th. garganica,* etc.) have the power of producing redness of the skin and vesication.

with the same care. This complication, which most often occurs later in the disease, may be cotemporary with the first pulmonary lesions, or even precede their occurrence for a longer or shorter time. Laryngitis, however slight in character, when occurring in a person threatened by tuberculosis, should never be neglected, since it may either indicate that a deposit of tubercle is about to take place in the larynx, or the irritation may extend thence to the trachea and bronchial tubes, and hasten the development of pulmonary changes. Thus laryngitis should be looked upon as a complication which may have the most injurious effect upon the ultimate issue of the disease, and the fundamental principle which I have so frequently enounced as to the harmful effect of respiratory irritation shows the necessity of immediate and continued intervention. Though the primary disease should not be disregarded, the laryngitis will require some special form of treatment, and on this account the inhalation of appropriate medical agents, local applications made by means of the laryngoscope, or such counter-irritants as croton oil, or iodine preparations applied to the laryngo-tracheal region, will be found beneficial. These simple means of treatment, when employed perseveringly, usually cure the early form of laryngitis which is now considered. It must never be forgotten that any form of treatment should be aided by as complete rest to the larynx as possible, and I have elsewhere mentioned the distinguished physician who, being affected with a suspicious form of laryngitis, was brave enough to remain for more than a year without speaking, a sacrifice which was followed by complete recovery. So much privation need scarcely be exacted from ordinary patients, but they should be made to understand the advantage of prudent restriction in speaking, a precaution specially indispensable when the profession is one which requires excessive use of the voice. Unfortunately, in such cases there is unusual difficulty in obtaining this result, but none the less should it be enjoined with unceasing perseverance. Knowing the evil effects of irritation, the physician must forbid any profession which exposes the patient to dust, or obliges him to inhale irritating or caustic vapours—a fact so evident that it need merely be mentioned.

In the preceding chapter the means of fulfilling the indication furnished by malnutrition and pulmonary inaction were considered, and now those will be discussed by which the effects of irritation and local lesions may be prevented. The combination of these means is required in practice, and the treatment, though complex, should invariably be employed in the primary apyretic stages of confirmed tuberculosis. No change should be made so long as the cough is but occasional and dry, but whenever habitual expectoration, however small in amount, shows the existence of pulmonary catarrh, this condition should be treated, since it not only adds to the fatigue and spoliation of the patient, but also favours the extension of tubercular lesions.

The importance, therefore, of this new indication is by no means inconsiderable, and for many years I have fulfilled it by means of balsams, tar, or turpentine, taken internally or in the form of inhalation. Since, however, the works of Bouchard and Gimbert have shown the good effects of the pure creasote yielded by the beech, these remedies have been abandoned, my own observations having confirmed the results obtained by my eminent colleagues, and shown me that creasote is more beneficial. Acting more rapidly and with greater certainty than other remedies, it diminishes the expectoration, and by its effect upon the bronchial tubes prevents any extension of the catarrhal lesions, thus notably reducing the extent of the pulmonary changes. Nor is this all, since creasote seems also to have some effect upon the fundamental lesions themselves, and to promote the sclerotic change by means of which recovery is found to occur in this disease.

I have often found in patients who were thus treated, that when the symptoms due to peri-tubercular catarrh have ceased to exist, and the extent over which stethoscopic signs of disease could be recognized was reduced to its minimum, a second diminution of the affected part took place two or three months later, accompanied in two cases by bronchial breathing and bronchophony at the periphery of the affected part. These signs, being accompanied by general improvement in the condition of the patient, seemed due to sclerosis in the part surrounding the lesion, which therefore limited the size of the softening focus. One of these cases was observed at the hospital, the patient being a female aged twenty-two, with deposit of tubercle at the apex of the left lung, which was now in the softening stage. During the three months that she remained, in addition to the ordinary treatment, creasote was employed, and at her departure not only was she found to have increased as much as fifteen pounds in weight, but the local changes already mentioned were found to have occurred. She remained in this improved state for almost two years, but in the intensely cold weather of December, 1879, was affected by bronchopneumonia, which caused her to come again under my care at the hospital in the most serious condition. Owing to some administrative difficulty, she was obliged to leave at the commencement of convalescence, and it is doubtful whether after this second attack she can ever return to the state of relative immunity from disease which she enjoyed for nearly two years.

The second case was that of a Russian aged thirty years, who owed to this plan of treatment the reduction of a softening focus at the apex of one of the lungs to less than half its size. It is now two years since this change occurred, and my last examination of the patient, six months ago, showed that the local improvement still remained, while the general condition was most satisfactory. The treatment being continued, more improvement may possibly occur, while in any case, on account of the hygienic conditions in which he lives, that which now exists may be looked upon as permanent, should no accidental disease affect the respiratory organs.

On account, then, of these and other analogous facts, creasote, combined with other means of treatment, seems to act not only upon the catarrhal, but also upon the tuberculous lesions, causing the sclerotic change to predominate over that due to sloughing or caseation. It may be seen that the whole plan of treatment which I adopt is with the view of obtaining this result, agents which improve the constitution, and the constant employment of alcohol, combining to effect this result. The fact that the latter has a tendency to promote sclerotic change is now definitely established.

Independently of these results, creasote may be also recognized to have an antiseptic or anti-putrid effect, on which account the pyrexia due to absorption (*la fièvre de résorption*), of which the importance will soon be recognized, is averted for a considerable time.

On all these accounts, therefore, creasote is a valuable remedy, and since my own observations have confirmed those of my colleague Bouchard, I never fail to employ it. From the moment that the special indication already mentioned is found to exist, creasote is administered. The indication furnished by the catarrhal condition is constantly in existence when tuberculosis has made the slightest progress, being only absent at the very first onset of the disease. This remedy is, therefore, one of the fundamental agents in the treatment which I adopt. It has also a permanent effect, and when once administered should be continued, since the indications which it fulfils are themselves persistent. Nor is it immediately that creasote produces its effects. Some days must pass before any good result is obtained, while pyrexia does not counter-indicate its use, the sole obstacle to its employment being the gastralgia which at times occurs when the remedy has been taken for any length of time.

Some plan should, therefore, be adopted which as far as possible prevents the occurrence of this complication. It is specially necessary on that account to employ a comparatively small dose. The remedy can be tolerated for months, in fact, as long as seems necessary when a small dose is administered, while it is at the same time beneficial owing to the continuance of its action. A large dose, however, may not be borne for more than a few days owing to the cause mentioned, and when once this happens, for any reason, it will be most difficult to again obtain its acceptance. It should, therefore, be the rule to administer a small dose at first, and to increase it but gradually, beginning with the amount of not more than three minims (*vingt centigrammes*) daily, which may be increased by one minim (*cinq centigrammes*) in each week or ten days until as much as five or even six minims are given. This dose should never be exceeded, and is but rarely reached, since, as has been already explained, a small dose is truly effectual, and does not expose the patient to the danger of gastralgia, which at any period of the disease is a real complication. From three to four minims daily is my usual dose, three or perhaps but two minims being administered when the remedy is first employed.

I have ceased to administer creasote alone, that is in the form of capsules, on account of the irritation which it produces in the mucous membrane of the stomach. I have also abandoned the employment of wine containing creasote, of which the taste is most disagreeable, and which soon produces invincible repugnance to its use. Owing to the inconvenience which attends the use of these preparations, I have adopted a plan which introduces no fresh complexity into the treatment, but enables the remedy to be tolerated for an indefinite length of time. I cause it to be taken with the cod-liver oil or glycerine administered to the patient, and with regard to the oil, its nauseous taste is found to be corrected by this addition, so that many patients who refuse to take it alone will readily do so when it is thus modified. Three minims of creasote may then be taken with the daily dose of oil, and if one drop of essence of peppermint is added, the fresh and penetrating taste of this aromatic herb will alone remain in the mouth. Thus, if the patient takes two ounces (50 *grammes*) of the oil, the prescription of the remedy for daily use would be as follows: Cod-liver oil, two ounces; creasote made from beechwood, three minims; essence of peppermint, one minim. One or more pints of this mixture could be easily kept in readiness, but care should be taken, when the dose of oil is increased, to modify the proportion of creasote in such a way that the quantity taken daily may not exceed five or six minims, which, as I have said, is my maximum dose. It is better, if possible, to make daily a fresh preparation of the remedy.

When glycerine is taken this precaution is less necessary, since the dose of the vehicle remains unchanged. After many trials, I have adopted the following prescription as the daily dose: Glycerine, twelve drachms (40 *grammes*); cognac or rum, three drachms (10 *grammes*); creasote, three to six minims; essence of peppermint, one minim;—this representing about four tablespoonfuls. The mixture is pleasant to take, causes no irritation of the stomach, and is invariably well tolerated even when taken for a long time; in this way it is preferable to the cod-liver oil mixture, and enables a most valuable remedy to be taken. At the same time, it must be repeated that every expedient should be tried to establish and continue the treatment by cod-liver oil.

I have repeatedly said that alcohol in some form should constantly form part of the remedies administered. Thus to patients who can tolerate oil, from one to two ounces of rum or brandy should be given daily, in the form of liqueur taken during meals or in the interval between them, and it has been explained why so much value is attached to this remedy.

In a similar way, the indication furnished by the strumous form of phthisis may be fulfilled by administering milk containing iodine, that is to say, when cod-liver oil cannot be borne in sufficient quantity. Glycerine is beneficial to the malnutrition which exists in all forms of phthisis; but in administering this remedy alone, the special indication resulting from the scrofulous origin of the disease will be neglected, to the great detriment of the patient, while milk containing iodine affords the means of fulfilling this important necessity without complicating the general plan of treatment.

The constant and variable indications furnished in the apyretic periods of the prolonged form of ordinary phthisis having been discussed, the febrile phases of the disease will be now considered.

Whilst the occurrence of pyrexia indicates a fresh stage in the progress of tuberculosis, it means also that some modification should be made in the therapeutics of the disease, since pyrexia should be always and perseveringly treated. Pyrexia is itself a cause of wasting, this being, as you know, one of my fundamental principles in treating the disease. Consequently, although here but a symptomatic phenomenon, it is no less necessary and urgent to oppose it. By removing pyrexia, a cause of rapid wasting is taken away, which, if left to itself, may hasten the consumption of the body, and lessen the duration of the interval which occurs in the fundamental treatment of the disease. Lastly, the functions of digestion are protected, and the production of dyspepsia, which but too often becomes permanent, is as far as possible prevented. Pyrexia should, therefore, be opposed, and this necessity is, in my opinion, one of the most urgent in the treatment of pulmonary phthisis.

That this may be done with the greatest benefit, and the effects of fever and condition of the patient be more exactly appreciated, the different causes of pyrexia in tuberculosis should be well understood. It would be a great mistake to attach to it always the same meaning, and to confine one's self in practice to the more or less barren and vague notion that pyrexia exists. In this case, more than in any, recourse must be had to the light thrown on the subject by pathological analysis, and I would recall briefly the fresh conclusions which observation has now led me to form, already discussed in the pathological portion of this work, and the fundamental types of the disease being alone considered.

At any moment of the complaint pyrexia may be due to the formation of primary or secondary granular deposits in the lung. Often the first symptom to declare itself, in by far the most cases it is intermittent and quotidian, the febrile attack taking place in the evening, with its three stages clearly defined. The cessation occurs during the night, some time before morning. Exceptionally the attack takes place in the early morning, and the lowest temperature of the twenty-four hours is found at the evening examination. This is a variety of inverse type which is frequent in acute miliary tuberculosis. The pyrexia may have a remittent character, being continuous but of much less intensity in the morning. This, however, rarely occurs except during the formation of secondary granulations, that is to say at an advanced period of the disease. On account of its origin it may be called the pyrexia of granulation (*fièvre de granulation*), or formation of tubercle.

At any moment of the disease fever may be again due to the development of pneumonic or broncho-pneumonic foci, either simple in character or due to the presence of tubercle.

Continuous and most intense in the evening, it lasts as long as the inflammation causing it, and differs but little in character from the pyrexia which occurs in primary broncho-pneumonia. It is in fact an inflammatory fever.

When the tubercular deposit begins to soften, other causes of pyrexia are added to those already named. It may be due to ulceration resulting in the more or less rapid formation of caverns. Continuous, and more intense at night as in the preceding form, though the remissions which occur in the morning are perhaps less in degree, no considerable or constant difference seems to exist between these two forms. At the same time, it is easy to distinguish them by means of the stethoscope, which causes very different sounds to be heard in the two cases. This form of pyrexia may be termed that of ulceration or excavation.

Lastly, from the time that the stage of softening commences until the end of the disease, including that in which cavities are formed, pyrexia may be connected with the absorption of products of destruction with which the lung is loaded, as well as with bronchial or cavernous secretions. These elements produce fever, being its most frequent cause, while the pyrexia is most serious on account of its duration, which may be indefinite if it be not checked. It may be of intermittent type, the attack occurring on the evening of each day; or of the double quotidian type, the first attack taking place at about eleven o'clock in the morning or at midday. Most commonly, however, it is remittent, being less intense in the morning, though the temperature does not become normal, a difference of from 2·7 to 3·6° F. (*d'un degré et demi à deux degrés C.*) often existing between the morning and evening temperature. In both the intermittent and remittent variety a rigor or attacks of shivering may occur at the commencement of the evening ascent, and the termination of the paroxysm is usually accompanied by abundant sweating. This form of fever, the pathology of which may be thus explained, corresponds to what has been vaguely designated hitherto the hectic fever of tuberculosis. In my opinion it is the fever of absorption.

Such being the widely different causes of fever in phthisis, the effects of treatment must necessarily vary in the same way. In the three first kinds pyrexia is the symptom of local action, lasting as long as that action continues. No exception seems to exist to this rule, except when in an early stage of the disease fever is connected with the first formation of miliary deposits. In that case the pyrexia may be checked for a longer or shorter time, though the deposit of miliary tubercles continues, as is clearly shown by means of the stethoscope. As regards the "absorption fever," which term I unhesitatingly substitute, with its above-mentioned characters, for the so-called hectic fever of authors, this indicates true septic infection. Without corresponding anatomical change, if uncomplicated, it may then be removed by appropriate treatment maintained or repeated with sufficient perseverance.

It might be inferred from what has been said that the latter form of fever need alone be opposed, and that the other kinds must be abandoned to the dominating influence of the loalc affection which produced them. This would be a serious mistake. It is true, with the exception already mentioned, that these forms of fever, which are symptomatic of an organic lesion, persist, whatever treatment be adopted, during the whole time that the morbid action continues. Though, however, their existence cannot be prevented, their intensity can be modified in all, or almost all, cases; and it is of no slight advantage to prevent, even in a slight degree, the rapid wasting which every febrile attack must occasion. It is for this reason that, notwithstanding the different effect which treatment may have, I employ it not only in the three above groups, but also in what I have termed the "fever of absorption."

TREATMENT OF THE ORDINARY FORM OF PHTHISIS. 165

Thus I usually treat symptomatic fever by preparations of quinine, though when inflammatory or ulcerative fever declares itself in a patient who has already suffered from that of absorption, I prefer giving the antipyretic remedy which I employ in the latter case, namely, salicylic acid, as will soon be explained. With this exception there is, in my opinion, no indication which should cause quinine to be abandoned. Being specially anxious with regard to the functions of the stomach, I administer not the sulphate, but the hydric bromate of quinine, which produces much less irritation to the mucous membrane of that organ. On the other hand, it is essential to fulfil two conditions, that the remedy may not be employed for a long time, and be really beneficial. To obtain these results, it should be administered in the way which ensures the greatest effect being produced by the dose given. This precaution is too often neglected, and by misjudging the amount which would be beneficial, intolerance is produced, and perhaps dyspepsia, without any pronounced modification in the febrile condition. Fifteen grains (*un gramme*) of sulphate, or twenty-two grains of hydric bromate of quinine, should be administered—a quantity of undoubted power when thus administered, but which, if divided into five or six doses taken during the twenty-four hours, has little or no effect upon the disease. The cause of this difference is the fact that the patient is at no time submitted to the effect of the whole dose, on account of the rapidity with which it is eliminated, so that when the hour arrives in the evening for the attack to take place, the paroxysm occurs as before, while the amount of from six to nine grains (40 *à* 60 *centigrammes*), which has been administered according to the plan adopted, is sufficient to prevent its recurrence.

My principle is as follows : to make the complete action of the quinine, as far as elimination will permit, coincide with the hour which precedes the increase or development of pyrexia. It is in the mean, after an interval of six hours, that quinine produces its full antipyretic effect, so that the remedy should be given seven hours before the onset of the paroxysm, the whole amount being taken within fifteen or twenty minutes at most. The plan which I myself adopt is the following: on the first day four tablets (*cachets*) are administered, at intervals of ten minutes, each of which contains seven and a half grains (*cinquante centigrammes*) of quinine, thirty grains (*deux grammes*) being thus taken by the patient within half an hour. If the temperature is not two degrees lower than that of the preceding evening, I administer on the following day twenty grains (*un gramme et demi*) in a similar way.

Whatever the evening temperature may have been, even if it does not exceed that of the morning, I administer fifteen grains (*un gramme*) on the third day, in two doses separated by an interval of ten minutes. No remedy is then employed during two, or more often three days, after which, if necessary, I readminister the quinine according to the same plan. I have employed this plan of treatment in a large number of cases, and, as I believe, without meeting with a single case of dyspepsia which could be referred to it. The pyrexia has invariably become of much less intensity, not only diminishing but often ceasing on the second or third day of active treatment; and during the days upon which the remedy is omitted, the recurrence, which occurs constantly unless the local action which produced the fever has ended, scarcely ever causes the temperature to rise to the same height as before.

The pyrexia due to the formation of tubercle, abnormal on account of the early date at which it occurs, and which, in patients who are specially excitable or weak, accompanies the first manifestations of disease in the lungs, is always much diminished, and in many cases removed, by employing the above plan of treatment twice or perhaps three times. Even one application may have this effect. The result is less satisfactory when the pyrexia is due to the formation of secondary miliary tubercles at a later date, near pre-existing tubercular or pneumonic foci. I have never known the pyrexia to cease so long as the local change is in progress; and though, after being remittent in character, it may become intermittent, this seems to be the sole change which occurs, though, on the other hand, the evening temperature becomes in almost all cases notably lower, often as much as 2° F., so that the wasting due to pyrexia is reduced to a minimum degree.

The results are appreciably the same in the inflammatory fever which attends attacks of congestion, intercurrent bronchitis, pneumonia, or broncho-pneumonia. My observations enable me to value the reduction of temperature which follows the first administration of hydric bromate of quinine at 2° F. (*un degré C.*); on the second day there is also a reduction of about 1° F., and this is maintained or perhaps increased after the smaller dose of the third day. During the two or three days which follow, upon which quinine is not administered, the evening temperature almost invariably remains lower than before the employment of the remedy, notwithstanding the elevation which most usually attends its suppression. When quinine is administered a second time, the temperature is again reduced beyond the point which it reached on the first occasion; and I have often seen the pyrexia cease, though the local lesion whose existence is thus indicated may be still in existence. The effect of the quinine is, therefore, quite as powerful as when antipyretic treatment is employed in acute pneumonia.

It is very different in the third form of fever, belonging to the group which is now considered. The fever produced by ulceration offers almost invincible resistance to all antipyretic remedies. No more can be obtained than a reduction of about

1° F. in the evening temperature, which during the interval nearly always attains its previous height, nor could I obtain a better result by altering the mode of administration, or varying the agents employed. This form of pyrexia is thus the most tenacious which occurs, and care must be taken not to attribute to the influence of the remedy a phenomenon which is at times observed in the same circumstances. When the tissue which surrounds the zone in which ulceration is taking place, whether containing tubercles or not, is free from inflammation, the pyrexia may entirely cease when the elimination of the diseased part is complete. This is, so to speak, a most favourable occurrence, though unfortunately but too rare. A high fever, which had been energetically but vainly opposed for several weeks, may be then seen to cease suddenly in the space of from forty-eight to sixty hours; and should this change coincide with the employment of any fresh remedy, there would be some inducement to ascribe to it the apyrexia, which is really due to the fact that the local affection has completed its existence, as already shown by modification of the stethoscopic signs and expectoration.

It should be repeated, in order to avoid a melancholy error, that pyrexia, due to the formation of a cavity, is the least influenced by therapeutic treatment, the causes which produce it being usually multiple in such cases. In addition to the ulcerative process, inflammatory action, or a deposit of miliary tubercles, is taking place around the diseased part, and usually also the absorption of putrid matter. Notwithstanding the complexity of the condition, the latter element may be supposed to exist, when the elevation of temperature is preceded and accompanied by shivering, which returns each day at nearly the same hour. When this occurs I cease to administer quinine, and have recourse to salicylic acid, which always forms part of my treatment when pyrexia due to absorption exists. Before considering this subject, however, some remarks should still be made with respect to the treatment by quinine.

Although hydric bromate of quinine is without doubt relatively innocuous to the mucous membrane of the stomach, I would not recommend its administration by the mouth to patients who have already suffered from dyspeptic troubles, or in whom, owing to a morbid condition of the appetite and digestive functions, such a complication is to be feared. If the quinine is to be administered during one or two days only, this precaution may be disregarded, but if the remedy is to be continued, this should be taken into consideration before it is done. By means of subcutaneous injection, the indication furnished by pyrexia may be wholly fulfilled without any irritation being produced in the digestive organs. In such a case the plan which I adopt is the following: If the hydric bromate is quite pure, fifteen grains (*un gramme*) will dissolve, without leaving any residue, in seventy-five minims (5 *grammes*) of distilled water. Fifteen minims of this solution, that is to say, the contents of the syringe of Pravaz (*une seringue de Pravaz*), will therefore include three grains of the quinine salt, which correspond to seven grains taken by the mouth. Consequently, by means of two or three injections, the effects of fourteen or twenty-one grains are respectively obtained. In conformity, therefore, with the principles which have been already explained, I make three injections at the same time on the first, two on the second, and one on the third day, a period of intermission then following. Since, however, the remedy acts more rapidly when administered subcutaneously, I operate five hours only before the onset of the attack.

These injections are not more painful than those of morphia. They produce no serious accident so long, of course, as the needle is introduced into the subcutaneous cellular tissue, and not into the deep layers of the skin. No sloughing of the skin, abscess, erysipelas, or glandular swelling ever occurs, as

I can affirm, after numerous observations. Around the puncture erythematous redness is developed over a space which rarely exceeds the size of a five-shilling piece, and which disappears in two or three days. It should be known, however, that a nodule remains in the connective tissue, which is indolent and adheres to the deep parts of the skin, and that this induration, which produces otherwise no inconvenience, persists in the mean from six weeks to two months. I have found traces of it after six months, but even then the nodule has in the end totally disappeared. On account of this peculiarity I make the injections as much as possible in the trunk or upper extremities, since in the lower extremities these nodules, when multiple, may be a cause of irritation to the patient when movements are made. With this exception their inconvenience is but slight, and may be totally disregarded.

I strongly recommend the employment of this process. When the injections were made of solutions containing an acid, and exposing the patient to serious inflammatory complications, it was natural that recourse should not be had to them without much hesitation, but the aqueous solution of hydric bromate of quinine causes no fear of such an accident, and offers a valuable resource which it would be a serious mistake to neglect.

The last form of fever, namely, that of absorption, may occur when softening begins to take place, but it more frequently does so at the time of real ulceration and the formation of cavities. At that time its occurrence is the rule, and pyrexia is constantly found to exist, though its continuity may vary. The special characters which this form of fever assumes have been already mentioned, which in fact enable it to be recognized, and in addition to this its comparatively late occurrence, and the stability of the stethoscopic signs as compared with those of inflammatory fever, are distinctive characters. Thus, if a patient in the second or third stage of phthisis is affected by pyrexia, and no change can be detected in the extent of the lesions either when it first occurs or on the succeeding days, the fever of absorption is known to exist by this fact alone. The pathological conditions, it should at once be stated, may not be ascertained without difficulty. Still, by means of the diagnostic characters already mentioned, and by considering the proportional degree of fever and the lesion which exists, the solution of this interesting pathological problem can always be obtained.

The fever of absorption is the most formidable kind which occurs, being typical of the phthisical condition with which it is connected. The pyrexia due to the formation of tubercles has a duration limited by that of the granular deposits, while that which is due to intercurrent inflammation or the formation of cavities terminates when the local affection ceases. The pyrexia of absorption, however, is different, and may continue indefinitely, that is so long as injurious particles exist which may be absorbed, namely, as long as the patient himself. Nor is this all; the fever is most harmful, not only increasing the emaciation, but producing true organic consumption, often the signal of incurable intestinal complications, and in all cases so completely modifying the condition of the patient that evidence of the approaching end is soon perceptible.

It need scarcely be remarked that this form of fever should be opposed with unremitting perseverance. Neither success nor failure should stand in the way; the treatment should be energetically and perseveringly employed, and by such means improvement may be obtained which did not at first seem possible. By following the above directions and adopting these principles, obstinate fever will often be found to give way, and a period of fresh improvement and relative comfort be thus assured to the patient.

Without doubt this form of fever may be successfully treated by quinine, and I would expressly recommend its employment whenever, for any reason, the remedy of which I am about to speak cannot be continued. When it can, however, I look upon salicylic acid as far preferable. On account, in fact, of the origin which I attribute to the so-called hectic fever, it should be opposed by an agent which joins powerful antiseptic action to its antipyretic effects. Such, in fact, is the double effect of salicylic acid. Thus, since first studying the effect of this remedy in 1876, I have employed it in the fever of absorption which accompanies phthisis, and my observations fully justified the opinion which I had formed. At about the same time and in the following years numerous physicians dwelt upon the antipyretic effects of salicylate of soda in the pyrexia of phthisis. A drachm of this remedy has a stronger antipyretic effect than fifteen grains of quinine sulphate, thirty grains acting in an equivalent manner. In my opinion, however, it is the antiseptic effect which causes its superiority in the fever due to absorption, and I find the proof of that special effect in the following fact. Upon treating this form of pyrexia by means of quinine, it improves, but if the remedy be omitted, fever reappears on the very day that this is done, and almost invariably with its original intensity. If, however, salicylic acid be employed, not only is the fever controlled, but absolute or relative apyrexia may continue for many days —perhaps a week or more—after cessation of the remedy. This cannot but mean that in the first case the cause which produces fever is not modified, its effects on the temperature being alone controlled, so that when the agent which reduces the temperature ceases to be employed the cause again acts with the same intensity as before. This cause, which in my opinion is the septic condition of the elements which fill the lung, would naturally now have the same power as before. In the case of salicylic acid, however, the very cause of the fever, namely, the fermentation of pathological products, is itself removed, and this being the case, the reduction of temperature persists even when its cause no longer acts, that is to say until new elements form in such abundance that by their absorption pyrexia is produced. Such is the reason why the time of apyrexia, which follows the omission of the remedy, is of such variable length. In other terms, it may be said that in the fever of absorption quinine alters the effects of a certain cause, while salicylic acid modifies both the effects and the cause itself—a distinction which may be thus expressed: in the fever produced by absorption quinine modifies the symptom, while salicylic acid tends to remove it.

My observations upon this subject are already most numerous, exceeding a hundred in number, and I have not seen a single case which disproves the theoretical and practical conclusions which have just been stated. The effects of salicylic acid are constant; if appropriate doses be employed with sufficient perseverance, the pyrexia due to absorption is invariably found to yield after a longer or shorter time. The reduction of temperature, which is at first gradual, becomes daily more pronounced, and if when this result is obtained the treatment be omitted, a certain length of time, at least three days according to my experience, will always pass before the fever returns to its previous intensity. So certain am I of the constancy of these phenomena, that I feel justified in naming them as a means of recognizing the character of the pyrexia. If the treatment be carried out according to rules which will soon be indicated, and neither complete nor temporary apyrexia occurs, the fever is not due to absorption alone, but is kept up by the formation of miliary tubercles, or by the occurrence of inflammation or ulceration. In such circumstances, though salicylic acid is not useless, it produces but partial and temporary results, as I have explained in speaking of the symptomatic fevers belonging to the first three groups.

In this discussion salicylic acid, and not salicylate of soda, has been considered. The truth is that numerous comparative observations have led me to believe in its superiority. As an antipyretic agent it certainly does not act more powerfully than the salicylate, but as an antiseptic its effect is more decided. Not that the salicylate has no antiseptic effect, but there is an undoubted difference between the two remedies to the advantage of the acid, not only because it acts more rapidly upon the pyrexia, but also on account of the longer duration of the apyrexia which it produces. It is for these reasons that salicylic acid is preferable, while the salicylate should be reserved for those cases in which intolerance prevents the acid from being used, this remedy being preferable to quinine. Thus when the fever of absorption is treated these three agents form a series of decreasing value in the following order: (1) Salicylic acid; (2) salicylate of soda; (3) hydric bromate of quinine.

When the daily ingestion of salicylic acid has to be tolerated during a length of time, two possibilities should be specially borne in mind, and regulate the employment of the treatment. In the first place, the dose given should be efficient, but not so large that cerebral complications might entirely counter-indicate any employment of the remedy after the first or second day; and secondly, care should be taken not to irritate the gastric mucous membrane. For these reasons I have adopted a new mode of procedure which is personal to myself.

To patients affected by the fever of absorption at an early period in the disease, and in whom a naturally strong constitution is not shattered by it, while the functions of digestion are unaffected, I administer thirty grains (*deux grammes*) of salicylic acid on the first day, and twenty or fifteen grains (*un gramme et demi ou un gramme*) on the second and third day, according to the case. If, after these three days, the fever has not abated, I again administer thirty grains (*deux grammes*), and pursue the same plan of treatment either at once

TREATMENT OF THE ORDINARY FORM OF PHTHISIS. 175

or after an interval. To patients in a less satisfactory condition I administer but twenty grains (*un gramme et demi*) on the first day, the same dose, or fifteen grains only, on the second day, and fifteen grains on the days which follow, continuing this dose or increasing it, after an interval of three days, according to the effect produced. In every case I continue the remedy at the dose tolerated so long as pyrexia exists, or at any rate while depression of the temperature is observed from day to day. The persistence of the fever is then due to a complex origin, and if the treatment be blindly continued, intolerance on the part of the digestive organs will undoubtedly ensue, or perhaps serious cerebral and cardiac complications. When this first stage of the treatment has produced its full effect, it may be discontinued, being recommenced should the pyrexia return in the least degree. At the same time, three full days of interval should be left even when the return of fever takes place more early, and by acting in this way the remedy may be employed with the same energy on the second occasion without signs of intolerance showing themselves. I continue to employ this treatment with caution and vigilance so long as fever exists, ensuring thus to the patient two important results: firstly, an abatement of the pyrexia should it exist; secondly, the repeated occurrence of apyretic periods, during which the phenomena of septic poisoning cease to occur, and febrile wasting of the organism no longer takes place. If the condition of the patient on account of this fever be considered, it would be difficult, in my opinion, to find a remedy more appropriate to the case.

The effect of salicylic acid upon the temperature of the body seems to be always more pronounced in proportion as the patient is more weak, or at a more advanced period of the disease. This fact should be taken specially into consideration when the amount of the remedy at first employed and the duration of the treatment have to be determined. Otherwise there is some danger of producing not only a lower temperature, but one of collapse; a fall of 7° to 9° F. (4° *ou* 5° C.) taking place, perhaps, in twenty-four hours, as I have twice observed, though fortunately without injurious consequence.

It should also be known how powerless are the effects of salicylic acid in patients affected by alcoholism, as I have frequently shown to be the case, having often met with proofs of this fact. I recently saw a clear example in a literary person who made no secret of his previous habits. At three different times thirty grains of this remedy produced as a maximum effect a reduction of 0·7° F. (*de quatre dixièmes de degré C.*), while often there was no effect whatever. Assured of this fact, I gave on two succeeding days twenty grains of hydric bromate of quinine, when the temperature fell in twenty-four hours from 102° F. (39° C.) to 99° F. (37·2° C.), rising again to the previous point when the remedy was discontinued. Thus it is useless to fatigue patients suffering from alcoholism with treatment which must be ineffectual; the powerlessness of this remedy being in reality a circumstance which aggravates the prognosis of the disease.

The plan which I adopt is to administer salicylic acid in tablets of seven grains, and in such a way that the whole dose shall be taken within the hour, if from thirty to forty-five grains are to be the dose, or within half an hour if the amount is less. The whole quantity should be taken as long as four hours before the time of the attack, since its effect on the temperature, which commences from thirty to forty minutes after ingestion, is only complete after three or four hours. With each tablet a large glass of water is taken, containing two or three teaspoonfuls (*deux ou trois cuillerées à café*) of brandy, this being administered with the double object of diminishing the local effect of the acid on the mucous membrane of the stomach, and of maintaining the excretion from the kidney, which salicylic acid checks in a notable degree. This precaution, which, in my opinion, is absolutely necessary, must on no account be omitted, though it may render the administration of the remedy in the short space of time mentioned an impossibility, the patient either not wishing or not being able to take at the same time so large a quantity of liquid. Should this happen, the amount is divided between a certain number of hours, care being taken to administer the last dose four hours before the commencement of the paroxysm. On account of the rapidity of its elimination, the antipyretic action is not so pronounced as when quinine is given, though this is less inconvenient than in that case, since the antiseptic effect is always proportional to the amount administered, whether it be taken in two, four, or six hours. This result is independent of the effects of accumulation, which, on the other hand, is indispensable when the maximum of antipyretic action is to be obtained. Thus, without compromising the result finally obtained, the amount of salicylic acid administered may be divided between a far larger number of hours than can be done in the case of quinine.

When the whole amount mentioned, namely, thirty grains, can be taken within the hour, it must be administered in its natural state; a solution of this quantity should contain at least an ounce and a half of brandy or rum, the absorption of which, in so short a time, would not be always without inconvenience. If, however, the administration of this amount can be spread over a longer period the difficulty ceases, and, if preferred by the patient, the remedy may be taken in a liquid form. After trying numerous plans, I have finally adopted the following combination: salicylic acid, thirty grains; rum or brandy, an ounce and a half; wine (*vin cordial*), four ounces. The taste is pleasant; the tartness of the acid is concealed as far as is possible by the taste of rum and wine, and by taking the precaution, always indispensable, of giving a large glass of pure water after each dose of the remedy, this plan can be adopted without producing intolerance. It has the advantage of associating the salicylic acid and alcohol in one combination, but the necessity of taking water always exists, and I, therefore, only employ the remedy in this way when tablets cannot be taken. The prescription which I have given is one of necessity, and cannot be altered if the clearness of the liquid is to be retained; the quantity of brandy is the smallest amount which will keep the acid in solution; and the wine (whichever kind be employed) is a necessary vehicle, since, if without changing the other ingredients an aqueous or mucilaginous liquid is put in its place, the fluid becomes instantly and unmistakably cloudy and totally unfit for use.

That no practical difficulties may be omitted, I ought to say that in some persons the mucous membrane of the stomach is so irritable that salicylic acid cannot possibly be tolerated. On the first or second day gastralgia and vomiting occur, and the remedy must be discontinued. The indication is not then to be disregarded, but salicylate of soda, which has a far less irritating effect upon the stomach, should be given in place of the acid. It would be employed in the same way, namely, with alternate periods of omission, and a dose of a drachm or a drachm and a half should be given on the first day according to the case, from forty-five to sixty grains on the second, from thirty to forty-five on the third, and so on.

In this form, however, it may be difficult to continue the remedy, and on this account I adopted a plan of doing so, the history of which is not without interest. Of this I would now speak, namely, the subcutaneous injection of salicylate of soda. My observations were made in 1880, when forty-six injections were thus made in the case of eleven patients under my care, nine men and two women suffering from the fever of absorption. In the six first injections, administered to three separate patients, I preferred using a very small amount of the remedy, namely, eleven grains (75 *centigrammes*) of the salt, which I introduced by means of two injections practised successively in the two arms. I thus obtained the same effect upon the temperature as by means of a drachm (4 *grammes*) taken by the mouth in the same circumstances; that is to say, a reduction of 2° or 3° F. (1° à 1·5° *C.*), which continued to exist for two days in two cases, for three days in the third. Strangely enough, in this series of observations the antipyretic effect did not manifest itself on the day upon which the injection was made, only becoming perceptible on the following morning, and attaining its maximum on the evening of the second day, that is to say thirty-six hours after the operation.

Why are these effects so late in showing themselves, and why is the action of this remedy so slow as compared with the striking rapidity with which injections of quinine affect the temperature? The cause, in my opinion, is the viscidity of the liquid. In order to have a sufficiently effectual preparation contained in one syringeful (*dans une seule seringue de Pravaz*), the solution must be so concentrated as to have a syrupy consistence, though sufficiently fluid to be able to pass through the syringe. When by raising a fold of skin and introducing the syringe at the base of this fold, so that the solution may pass at once and undoubtedly into the connective tissue, the liquid is injected beneath the skin, it is observed that whatever care be taken the liquid accumulates at the point which corresponds to the extremity of the syringe, and there forms a small immovable elevation which is quite visible to the naked eye. In the six first injections mentioned above it was easy to recognize, both by the size of the enlargement and the absence of any swelling in the periphery, that the whole of the liquid had collected in such a way. In another patient in whom two injections were made by the usual method in the upper extremities slightly on the inner side and above the insertion of the Deltoid muscle, while in the left arm the whole of the liquid accumulated at one spot, in the right not more than half did so, while the rest spread into the surrounding cellular tissue. Since the amount injected was in both cases the same, the difference could be appreciated by difference in the size of the two subcutaneous nodules. Some variety may, therefore, exist in the local effect of the injection depending probably upon the different relation which the extremity of the cannula may have with the thick bundles of subcutaneous connective tissue. The above fact justifies this idea, for there was no difference in the two injections but in the direction of the instrument. In the left arm, in which the whole liquid accumulated together, the cannula was introduced perpendicularly beneath the skin from above downwards, while in the right arm it was introduced from below upwards, having an ascending direction. One thing is certain, that owing to its viscidity the liquid adheres either wholly or in part to the meshes of connective tissue into which it is injected, while the molecules themselves also adhere together for the same reason, conditions which end in the formation of a swelling which distinguishes this injection from all others that I have hitherto observed.

This being the case, absorption takes place in the little mass by very slow degrees, and so much the more slowly that all the liquid injected is contained in it, and the unexpected delay in the effect of the remedy should, in my opinion, be attributed to this cause. Within twenty-four hours from the time of the injection, the swelling is much smaller, though remaining of about half its previous size; and it is only in from thirty-six to eighty-four hours from the time of the operation that the work of absorption may be looked upon as ended. At this time a nodule of insignificant size alone remains, adherent to the deeper layers of the skin, and resembling those determined by the injection of hydric bromate of quinine, except that it is much smaller in size. In one or two days this last sign of the injection disappears, so that on the tenth day no evidence whatever remains.

The solution which I employ is composed of equal parts of salicylate of soda and distilled water. I have ascertained that the quantity of liquid contained in the syringe of Pravaz weighs twenty-two grains (1·50 *grammes*), so that each injection would contain eleven grains of the salt. At first it seemed advisable to inject half the liquid into each arm, but when this practice was found to be undoubtedly harmless, I injected the whole quantity into one arm, so that an amount of the salicylate equal to twenty-two grains presented itself for absorption. The effects are more pronounced, more rapid, and more persistent. The solution should be prepared in small quantities at a time and kept in a bottle of blue glass, since its colour will otherwise become of a darker hue, though without other change taking place.

When I had once commenced to inject twenty-two grains of salicylate at the same time, delay in the reduction of the temperature was no longer observed. This became always perceptible on the evening which followed the time of the injection. Though I could perceive, as when the injection was made, that a small swelling existed on account of the total or partial accumulation of liquid in one place, still, owing to the increased dose, absorption could take place during the time comprised between the morning and evening observations, and in sufficient quantity to produce reduction in the temperature. No other explanation, in my opinion, can be given of this interesting difference.

On account of the extreme solubility of the salt, one might be tempted to use a more concentrated solution than the above. This, however, is unadvisable for many reasons; the liquid becomes so viscous that it can scarcely flow through the cannula, the pain caused by the injection is most severe, there are signs of inflammatory reaction being set up, and lastly the solution does not remain clear, part of the salt being then precipitated. On all these accounts the above-mentioned solution is, in my opinion, the strongest which should be employed.

A few words will now be said about the local effects of the injections. These are perfectly harmless as far as consecutive inflammatory consequences are concerned, so long as the injections are made according to the rules already laid down, and the point of the needle is at no moment in contact with the deep layer of the skin. If this requirement is properly fulfilled, the local reaction is quite insignificant; a diffused redness around the point of insertion, which disappears in two or three days, pain due to injection of liquid into the areolar tissue, which is more acute than in the case of morphia or distilled water, and during the next two or three days a disagreeable rather than painful sensation is provoked by pressure or movement at the seat of the above-mentioned swelling.

Since in the forty-one first injections practised nothing occurred of more consequence than what has been stated, had I then concluded my observations there would be nothing to add to the details which precede. In the three following observations, however, sloughing took place, which affected the whole thickness of the skin, and took some time to heal. The only cause to which, in my opinion, this accident can be attributed is penetration of the needle into the thickness of the corium, the result of some sudden movement on the part of the patient. These three individuals, in fact, two men and a woman, had already received injections of the salicylate without any consecutive accident. The solution and instrument used were the same, the injections were made in precisely the same way as the others, and the necessary precautions in raising the skin were undoubtedly taken. Lastly—and this seems to decide the question—while the injection into the right arm of the woman was followed by sloughing, that made at the same time into the left arm produced no bad consequence. Two days later injections were twice made with the same liquid and the same syringe, and without the occurrence of any complication.

The sloughing, therefore, which was observed in three out of forty-six cases must have been an accident, due to the operation itself. The character of the liquid cannot have been the cause; but these facts show how necessary it is to take every precaution in making injections, and on account of the accidental nature of the complication, which could not have been foreseen, to reserve this form of treatment for cases in which salicylate of soda cannot be given by the mouth, it being then necessary to employ the remedy in this way.

The antipyretic effect of subcutaneous injections may be considered undoubted, and a valuable resource against the pernicious effects of fever. The advantage of this treatment is not naturally limited to the pyrexia of phthisis, and if the above-mentioned precautions be taken, recourse may be had to it with the same advantage in all diseases in which the salicylate of soda can be employed.

The minute details into which I have entered with respect to different antipyretic remedies, and the care with which their effect has been studied, show what great importance I attach to antipyretic treatment in the disease. To employ it is the fundamental duty of the medical attendant, which may have to be fulfilled at any moment. The pyrexia must be treated in any case, and at any time. Whilst it continues no improvement is possible, and on this account alone the condition of the patient becomes daily more serious. The termination of the pyrexia may be the signal of an improvement, which may be of some duration. Even if the pyrexia is only diminished, a great service is still done. To be energetic and persevering is the rule that should be in force, but in order to make the means of treatment adopted more beneficial, the distinctions between the different origins of pyrexia already mentioned should always be borne in mind.

CHAPTER VIII.

TREATMENT OF THE ORDINARY FORM OF PHTHISIS (*Continued*).

Modifications of the treatment during the stages of pyrexia—Hygiene—Diet—Medical agents.
Indication of antimony—Mode of administration and effects—Indication of quinine after the employment of antimony—Treatment of intercurrent attacks of acute pneumonia or broncho-pneumonia.
Treatment during the formation of caverns—Clinical distinctions—Desiccation of caverns and means of promoting its occurrence—Remedies taken internally—Inhalation of carbolic acid—Indication, effects and mode of employment—Antiseptic atmospheres.
Treatment of gastro-intestinal disorders—Catarrh of the stomach and intestine—Dyspepsia—Vomiting—Diarrhœa.
Treatment of hæmoptysis—Hæmoptysis without or with pyrexia—Late hæmoptysis.

THE employment of antipyretic remedies is not the only alteration in the treatment already adopted which is required by the occurrence of pyrexia, and from this point of view the distinctions mentioned regarding the origin and signification of the fever will be found of practical importance, since the modifications which should be effected in the general plan of treatment are by no means the same in different cases.

The various kinds of fever will, therefore, be successively considered, and the modifications which should or should not be made in the fundamental treatment adopted, without, however, any repetition being made of the treatment required by the pyrexia itself, which has already been considered in detail.

The pyrexia which accompanies the first formation of

tubercles is at times attended by dyspepsia of such severity as to interfere with the nutrition of the patient. When this occurs, the diet must be restricted and adapted to the diminished power of digestion, the milk, meat in its most simple forms, and wine being if possible administered as before. With regard to external life no change should be made; in the large majority of cases the febrile attacks take place in the evening, and terminate at an early hour of the night, as already explained, this supplying no reason for any change in the habitual occupations of the morning or afternoon. It is, in my opinion, a serious mistake to look upon these attacks as indicating that the patient should remain in bed, or even in the house. Doing this would be likely to hasten the onset of dyspepsia, and to prolong the duration of the fever. When, as exceptionally happens, the early fever of phthisis is of the remittent type, it may be advisable, if the morning temperature exceeds 100·5° F. (38° C.), for the patient to remain in his room, though not in bed, until the treatment adopted has caused the pyrexia to become intermittent in character. Except in this uncommon case, every rule as to life in the open air, walks and exercise, should be observed as before. The same may be said of hydropathy; if the patient has already submitted for some length of time to the habits required by it, there is no reason to omit a practice which offers at this moment the advantage of promoting the effects of antipyretic treatment. At times, however, it is necessary to yield to prejudice and fear, in which case the shower-bath (*la douche*) should be replaced by cold lotions applied in the morning by means of a large sponge, and followed by dry and strong friction as already explained.

It is evident that the time at which the febrile attack occurs should not be chosen for aërotherapeutic practices, which with this reserve should undoubtedly be continued or instituted, observation having shown that lesions which were but of slight degree before this is done, notably catarrh of the apices, may improve or even cease, owing to the effect of this treatment. Since it is without effect upon pyrexia, no reason exists which can justify its exclusion in the circumstances which are now being considered. The observations made by Forlanini by means of movable machines have shown that, even while fever exists, this practice does not raise the temperature as much as 0·2° F. (0·1° C.), and that this elevation does not continue for more than four minutes after the inhalation has been made. This effect, which is not perceived during the apyretic periods of the disease, may therefore be totally disregarded.

The medical treatment should be continued or adopted in its entirety, care being taken, however, to substitute glycerine for cod-liver oil whenever any doubt exists as to the integrity of the digestive functions. In adapting the treatment to meet the new condition of things consequent on the occurrence of fever, there is naturally an infinite number of individual distinctions which cannot be foreseen, and it has been my object to explain the general principles according to which the treatment should be directed, rather than those adapted to individual cases.

In the febrile stages due to congestion or inflammation the requirements are altogether different. Whatever anatomical lesion may be the cause, this incident constitutes an acute complaint, which transforms the person into a true invalid in the complete sense of the word. Confinement to the room, or often permanently to bed, is absolutely necessary, and on this account all the hygienic and mechanical means of treatment are temporarily suspended. The diet must naturally be restricted in proportion to the individual toleration and intensity of the acute condition, though however serious the latter may be, too low a diet should never be prescribed. In all cases broths, soups, the gravy or jelly made from meat, and above all milk, should be taken as nourishment, the latter being continued so long as it does not provoke distaste or diarrhœa, or fatigue the stomach. On the other hand, koumiss, if previously taken, should be discontinued, since observation has shown it to be most unsuitable when either intermittent or remittent fever exists. On the whole, then, the rules as to the diet of the patient are the same as those usually made in acute disease. Nourishment must be administered, and in this case the necessity is specially imperious, and should be fulfilled as far as possible on account of the wasting effect of the disease. These complications, even when they end in resolution, and cease without any increase of the pre-existing lesions, always aggravate for some length of time the condition of the patient, not only because of the pyrexia which they induce, but also, and more specially, on account of the interruption which they cause in the hygienic practices, nourishment, and often also in the medical treatment of the disease.

With respect to the latter point, the measures which should be adopted vary according to the intensity and character of the acute affection. If the pyrexia is not intense, if the congestion or inflammatory irritation of the bronchial tubes be but slight in character, it is not necessary to discontinue or even to modify the treatment already instituted. Cod-liver oil may be replaced by glycerine, and creasote may be continued if already taken; in short, the adjunction of antipyretic remedies is the only therapeutic change which, in my opinion, is necessary.

When, however, the symptoms, although of this nature, are more pronounced in character, when the fever is high and the morning remission slight, when the congestion gives rise to hæmoptysis, and the bronchitis, although but a temporary complication, affects the smaller bronchial tubes, and produces continuous dyspnœa, then such remedies as arsenic, cod-liver oil, glycerine, and creasote should be omitted, not because of their being injurious in these conditions, but from the necessity of employing a more active remedy against the tendency to congestion, which is the origin of pyrexia and other complications. Antimonial remedies fulfil this temporary requirement. The preference should, in my opinion, be given to mineral kermes * (*kermès*), administered in the dose of from two to five grains (15 *à* 30 *centigrammes*) daily according to the case, or the white oxide of antimony (antimonious oxide, Sb_2O_3), which, however, must be employed in the larger dose of from twenty to thirty grains (1 *gramme et demi à* 2 *grammes*) in the twenty-four hours, and seems to be less effective either in producing counter-irritation or as an expectorant. It may, however, advantageously replace kermes when the latter begins to produce nausea and vomiting, even in small doses, while its employment is still indicated. It is in the same group of cases, namely, when peritubercular congestion or acute bronchial catarrh exists, that tartar emetic (tartarated antimony), employed in small doses, may be also of real service. From three-quarters of a grain to a grain and a half should be administered daily in four or five ounces (120 *à* 130 *grammes*) of syrupy liquid given in table-spoonfuls. Its nauseating effect should be avoided, to effect which one or two table-spoonfuls of syrup of poppy (*sirop diacode*) should be added to the remedy during the two or three first days of its administration. This addition is then usually unnecessary, since when tartarated antimony (*tartre stibié*) is thus employed habitually toleration is apt to ensue. Whatever antimonial preparation is used, the digestive functions should invariably be watched with care; and if diarrhœa occurs it may be treated by opiates for a day or two, and should it persist after the use of this remedy, the treatment must be immediately and unhesitatingly abandoned.

* "Kermes is not a definite compound, but a variable mixture of antimonious sesquisulphide and sesquioxide, the latter being combined with a small portion of potassic carbonate or caustic potash." See Miller's "Elements of Chemistry," vol. ii.

When the remedy is well borne, it should be continued so long as the complication lasts which caused it to be administered. Whilst such is the case, recourse would not be had to quinine, firstly because antimonial remedies, even in small doses, have a pronounced effect upon pyrexia when their administration is continued; and secondly because such a combination would constitute a confused mode of treatment, which should be always avoided if possible. On the other hand, when these agents have produced their effect, quinine is certainly indicated; that is, if pyrexia continues when the complication which had been its cause has ceased. Such a case is not rare, the patients affected with this disease being always liable to attacks of pyrexia, which the least cause suffices to produce, and when once this happens the fever persists so long as it is not directly and energetically opposed. Thus, when some local complication has ceased to exist, the patient often continues to have intermittent fever, which declares itself in the evening, and no longer indicates congestion or catarrh, but the existence of tubercles which so far had been absent. The pyrexia is due to the acute complication, and to the febrile tendency which exists in the patient, so that this fever continues, notwithstanding that the complication which immediately produced it may have ceased to be the cause. In such cases, which are by no means rare, reliance should not be placed upon preparations of antimony, which are now valueless, but upon the so-called antipyretic remedies, and the pyrexia should be opposed by the means and processes already considered.

The form of pyrexia, which I have called, for convenience of language, inflammatory fever, may be connected with a more serious local affection, as with the occurrence of pneumonia, broncho-pneumonia, or miliary tubercles. In such a case the treatment should, in my opinion, be modified. Antimonial preparations which seem then to do more harm than good should be replaced by hydric bromate of quinine and alcohol, of which from one to two ounces (*de* 40 *à* 60 *grammes*) should be taken daily according to the case, either in a simple mixture containing syrup, or preferably some stimulating drink containing extract of cinchona. At the same time, if the difficulty in breathing is intense, if the pulmonary lesions produce stasis of blood in the healthy portions of the lungs, dry cupping should be practised, from forty to sixty cupping-glasses being applied to the lower extremities, and over the bases of the lungs. The patient should be nourished if possible by means of milk, broth, and raw meat; in short, the treatment adopted should be the same as at the onset of tuberculous pneumonia, since pneumonia, broncho-pneumonia, or acute granulosis may produce during the progress of ordinary phthisis the very condition which constantly exists in the pneumonic form.

In every variety of pyrexia, whether congestive or inflammatory, counter-irritants should be perseveringly employed. In the same way as fever and hyperæmia are treated by means of internal remedies, so must the local lesions be unceasingly opposed, in order to limit and terminate those which can end in resolution, every effort being made to prevent the acute complication from adding to the already existing disorders. Of the different modes of practising counter-irritation already considered, the successive application of blisters seems preferable to the others. In many cases it may be useful to provoke and maintain suppuration at the surface—a practice undoubtedly painful; but when the local lesions are recent, large in size, stationary, and threaten to remain in the acute condition, such treatment is more beneficial than counter-irritation of a less energetic or prolonged character. Upon this subject I again speak from experience, and cannot sanction the discredit into which this mode of treatment has fallen, no absolute counter-indication existing to its use except intense constitutional debility.

When the complication, unfortunately so liable to return, has subsided, the treatment previously adopted should be cautiously recommenced in its entirety. Even the hydrotherapeutic practices should be employed, in the form of a fine shower-bath borne for a short time if the necessary arrangements exist in the house inhabited by the patient, or in some adjoining building connected with it by a covered passage, or in the form of lotions applied in the room if the conditions are less favourable. Either the shower-bath or the lotion should be followed by dry friction, and if the strength permits, reaction may be obtained by means of a walk. Should this be unavailing, or should the length of walk necessary to re-establish the proper degree of warmth be too fatiguing, the patient should be enjoined to remain in bed for three-quarters of an hour or an hour, and it is most rare that before the end of that time the reaction is not completely and permanently established. In this point, as in many others, numerous individual varieties exist which prevent any general rule from being applicable. One rule should, in my opinion, be constant and invariable, namely, to forbid all aërotherapeutic practices in the case of those who have suffered, though it may be but once, from one of the complications already mentioned. The congestion or inflammation which attends their onset modifies the physical condition of the pulmonary vessels and of the lung itself in such a way that these practices are specially formidable as respects hæmoptysis.

The so-called "fever of absorption," at whatever period it occurs, imposes but little change in the treatment previously employed, so long as the digestive functions remain unaffected. As at all times of pyrexia, cod-liver oil is to be replaced by glycerine; but creasote should be perseveringly administered, the antiseptic properties of which are now doubly advantageous, and which forms a true auxiliary to the antipyretic effect of salicylic acid, which should be employed in these circumstances

Salicylic acid is similarly indicated in the fever which attends the formation of caverns; since if this fever, of complex origin, is on the one hand the expression of the ulcerative action and inflammation which attends removal of the diseased part, on the other it is as certainly the effect of auto-infection, the conditions being specially favourable to the absorption of diseased products. It is the more necessary to act with energy upon this form of fever, and upon the septic condition which exists, since this period in which septic absorption and excavation occur must on no account be considered as invariably the last stage of the disease. With regard to this fact, which is but too little recognized, and which has already been mentioned and dwelt upon, some important distinctions should be made. Without doubt, when the tuberculous deposits scattered through the lungs undergo at the same time destructive softening; when, from the double effect of sloughing and elimination, the deposits ulcerate, and break up on all sides, carrying with them the intervening pulmonary tissue, such a condition must naturally form the terminal stage both of the patient and the disease. On account of the simultaneous formation and extent of the caverns, whether of large or small size, not only is the lesion beyond recovery, but also unable to be of long continuance, since the hæmatosis must be insufficient in amount owing to the small proportion of healthy lung. Thus the patient is overcome as much by slow asphyxia as by the putrid absorption which takes place in the ragged walls of the caverns. The most favourable event which could happen in these conditions is abatement of the fever, and replacement of this period of acute destruction by a state of relative torpor. The wasting is then less pronounced, the need of hæmatosis less intense, and, owing to the influence of habit, the patient can lead a vegetative existence even though respiration is insufficiently performed. If, at the same time, the absorption of putrid matter is arrested, as shown by the cessation of fever, then, notwithstanding the pulmonary destruction, the patient may survive for weeks, or even months, since, escaping the causes of rapid death which have been indicated, he finally yields but slowly to the marasmus produced by persistent organic wasting.

This delay of the fatal termination, which in the above conditions is the most that can be expected, would not be due to natural causes alone. Therapeutic inertness must be avoided, and treatment should be actively employed both to abate the fever and to oppose the absorption of putrid products. By such means the strength is supported, the destructive action made to cease, and a torpid period established, resembling the calm after a storm. Alcohol, as restorative a food as the condition permits, and salicylic acid, or, in default of this, salicylate of soda, are the remedies which best fulfil these vital requirements.

Thus it is that even in the most formidable period of the stage of excavation, and in the most desperate conditions, remedies are not necessarily useless. The disease does not always progress in this precipitate and fatal manner, and the advance may be made but slowly in this sense that tuberculous infiltrations and deposits only occur at long intervals of time.

The danger which results from want of hæmatosis, owing to a portion of the respiratory surface being suddenly unable to act, no longer exists, the effects of absorption are less pronounced, and the general condition of the patient is more satisfactory and more amenable to treatment, which should be the same as that previously employed. During the intervals which exist between the different steps of ulceration, the state is no more serious than before, while the fundamental treatment employed may be to some extent beneficial. Alcohol, on account of the way in which it stimulates the organism, and leads to the formation of fibrous tissue, as already stated, may occupy a more important place than at an earlier date, and creasote should be still given, which not only arrests excessive secretion, but also has an antiseptic effect. It should never be forgotten that caverns may be closed and heal, this fact not being a clinical supposition, but one proved by pathological anatomy, as already shown by remarkable instances; in my opinion, the treatment which has been recommended contributes much to this fortunate issue.

Lastly, a result of excavation should be mentioned, which is most favourable in its effect, but more rarely observed. Destructive softening may take place at an early period in the disease, commencing before the organic wasting is yet pronounced, while the lesions are unilateral and limited to one deposit, which removes by destruction of tissue those injurious elements by which the part was infiltrated. The process of excavation is really one of elimination, ending in the formation of a cavern which, notwithstanding its size, may become smaller and eventually close. This closure may be the signal of recovery, so long as the tuberculous diathesis, exhausted by this manifestation, does not still produce other tuberculous deposits. Even when such a fortunate result does not occur, the formation of caverns at an early date should not lead to discouragement, and still less to inaction. In this limited and rapid form of the disease, the existence of caverns should be regarded as an incident which may be of a favourable nature, and every effort should be made to ensure to the patient the benefit of this possibility. All should, therefore, be done with the view of enabling him to weather the temporary storm produced by ulceration, it being remembered that whatever treatment be adopted, fever will exist so long as the elimination continues, all that can be done being to lessen its intensity. When this is complete and a cavern has been formed, the treatment adopted should have in view its desiccation and closure. In this case the repetition of counter-irritant applications has an important effect, and their employment should never be omitted.

In any case, so long as the caverns are neither so large nor so numerous as to make recovery impossible, every effort should be made to modify the liquids contained in them; not only owing to the danger of infection, but also from the fact that they irritate the adjoining parts so as to increase the loss of substance which exists, and but too often transform a cavity of fixed size into an extensive ulceration. In many cases the combined use of creasote and preparations containing salicin or salicylic acid will have this effect, while at other times the treatment in such doses as can be continued remains inoperative, the quantity of expectoration which indicates the amount of secretion in the cavity remains unchanged, the quality of the liquid is the same, and examination of the chest shows that, notwithstanding the expectoration, the caverns always contain more or less the same quantity, and continue to secrete liquid.

In such circumstances the physician is not merely to lament the inefficacy of his treatment, but to increase its action by the addition of agents having a more direct effect, as by employing the inhalation of atomized medicinal fluids. Different substances may be employed, specially iodine, tar, and turpentine, though in my opinion carbolic acid is to be preferred, being administered in the following manner.

A solution containing one per cent. of acid being employed, four ounces (100 *grammes*) are inhaled at first daily in the course of three or four sittings; the quantity inhaled is then increased without change in the strength of the solution so long as the treatment is well borne by the patient. This process may be carried out by means of the ordinary spray-producers, which are not worked by means of steam. For more than a year, however, I have employed the spray-producer of Lucas-Champonnière, in which steam is used, and have been, therefore, obliged to increase the strength of the solution employed, since when that apparatus is used the same quantity of steam is produced as of medicinal spray. A solution containing two per cent. of acid is therefore employed, of which two ounces (50 *grammes*) are placed in the bottle; to this two ounces (50 *grammes*) of steam are added when the spray is produced, so that when the remedy is used the patient inhales, as in the other case, fifteen grains (*un gramme*) of acid in four ounces (100 *grammes*) of water. The dose may be increased as easily as with the ordinary spray-producers, since if four ounces (100 *grammes*) of the two per cent. solution are employed daily, thirty grains (2 *grammes*) of acid would be combined with eight ounces (200 *grammes*) of water (on account of the steam mixed with it), which should always be the proportion at first used, so that the power of the treatment will have been increased without any change in the strength of the solution. I look upon this fact as important, because experience has shown me that a two per cent. solution is most beneficial to patients who are practising inhalation frequently during the day, and for many days continuously. It is often needful to commence with a solution containing but one-half or one per cent. of the acid. To diminish its effect upon the surface of the bronchial tubes, while the same amount of the remedy is taken daily, four ounces (100 *grammes*) of a one per cent. solution could be used. When the inhalations practised during the day are concluded, the patient will have inhaled fifteen grains (1 *gramme*) of acid, but this will have been distributed through eight ounces (200 *grammes*) of water on account of the steam added. In some cases I have employed during some months a four per cent. solution which, on account of the added steam, has the strength of one containing but two per cent. The patient inhales as much as ten ounces (250 *grammes*) of the solution daily in two sittings, of which each has a duration of from half to three-quarters of an hour. It seems doubtless that, notwithstanding the addition of steam, no sensation of heat is produced in the lining membrane either of the mouth or pharynx.

By means of three or four sittings, the desired quantity of the solution can be inhaled during the twenty-four hours. The patient is placed in a convenient seat, so that the column of steam corresponds with the orifice of the open mouth, and so near the instrument used that the vapour may enter the mouth before being diffused in the form of a cone whose base is at the periphery. If this precaution is not taken, the effect of half the liquid at least is lost, since the base of the cone exceeds in size that of the buccal orifice. Upon the patient making deep inspirations the fits of coughing soon show that the vapour has penetrated wthin the bronchial tubes, and to make that penetration more certain the patient may keep his tongue outside the mouth by means of a compress, though this, which is a most painful process, makes the treatment far more irksome, and seems quite unnecessary in the majority of cases. Inhalation, when effectively practised, is always accompanied at first, as has been already said, by fits of coughing, attended by more or less expectoration. These fits, however, take place at longer intervals after a few seconds, and the latter part of the sitting, which continues, it may be, for fifteen or twenty minutes, is passed without their occurrence.

On the two or three first days these inhalations produce some fatigue, and at times a sensation of giddiness; if these effects are pronounced, some alteration in the amount of remedy used should be gradually made, but half the quantity, for instance, being absorbed during the day. The patient, however, becomes rapidly accustomed to the practice, and being more ready to believe in a remedy which acts immediately upon the diseased part, submits readily to its use, notwithstanding the inconvenience which it undoubtedly occasions. One fact should be well understood, namely, that this form of treatment should never be employed when excavation is taking place, but after the termination of this more or less acute process, and when internal treatment produces no appreciable modification in the ulcerated surfaces.

If the above precautions are taken inhalations will usually be well borne, so long as care is taken to make the strength of the solution proportional to the irritability of the buccal and pharyngeal mucous membrane; experience, however, has shown me that a small number of patients exist who cannot be brought to tolerate this practice, namely, those who are in a state of profound debility, and who, though not feverish, are always liable to nervo-vascular irritability. Inhalation increases this condition, exhausts the strength, and may produce giddiness and liability to syncope. When this is the case such treatment should not be practised, the inconvenience attending its employment being of far more importance than the remote advantages which may be expected. The same reserve must be made when vomiting is produced, as not rarely happens in patients who are liable to this inconvenience during fits of coughing. In all these cases perseverance in the employment of the remedy would be a source of danger, and the practice should always be, at any rate at first, a mere therapeutic trial, since the observation of some days can alone show whether it is truly beneficial.

The diffusion of the liquid column in a conical form has already been mentioned, as well as the loss which results from the position of the patient; in any case, indeed, some portion of the inhalation is ineffective, the whole of the remedy employed being never absorbed. In order to diminish the loss, and moderate the force of the column which strikes the pharynx, I have had finer tubes applied to the steam spray-producer; this succeeds but imperfectly, since, though the column is of smaller size, it still exists, so that the face and clothes are moistened by the liquid. If this inconveniences the patient too much, it may be prevented by the employment of a conical tube with the large extremity adapted to the patient's mouth, while the other receives the column of vapour as it leaves the apparatus, and before any diffusion has taken place. The loss is not diminished, since one part of the liquid is deposited on the walls of the tube, but the inconvenience which has been mentioned now ceases.

When the practice can be satisfactorily carried out, as usually happens in eight or ten days, a pronounced diminution of the expectoration occurs. Any existing fetor will have ceased to exist; the symptoms which most plainly indicate absorption of putrid matter, such as anorexia, dyspepsia, and fever, which have resisted ordinary treatment, will have diminished, and the general condition of the patient have improved. These first effects of the treatment are usual, and almost constant, though in many cases nothing more is obtained even when the inhalations are continued for months without their employment being in any way interrupted. On the other hand, in some cases the amount of liquid secreted becomes less in amount, the gurgling within the cavities is diminished, and after some weeks, six or eight it may be, the secreting cavern, always on this account prone to extend, is transformed into a dry cavity of fixed size. From this time the patient may profit by everything advantageous which occurs in the local process, such as the *statu quo*, which has in itself a favourable meaning, and in exceptionally fortunate cases diminished ulceration and cicatrization. Thus the advantage which may be derived from adopting the whole treatment which I recommend can be appreciated. Though no perceptible effect is produced upon the tuberculous affection, the ulcerations and secretions in the bronchial tubes are so modified as to be in the best conditions with regard to recovery, so long as the organism of the patient and form of the disease render that fortunate result possible, while by improving the general condition of the patient it also tends to produce this effect.

How much of this result would be due to inhalation it is impossible to say, since I am so convinced that alcohol and creasote have a similar effect that inhalation has never been employed alone; at the same time, in some cases when internal treatment continued for some weeks had produced no effect upon the secretions of the cavity, and inhalations are then combined with them, the desired improvement is produced.

Cases may also be seen of similar lesions in which this local treatment is unable to produce any change in the quantity or quality of the secretion, and it is for this reason that the above-mentioned effects, the least to be aimed at, are not constantly obtained. If, after being carefully employed for twenty or thirty days, this plan of treatment leaves the condition of the parts unaltered, it should be looked upon as condemned, and any longer continuance of the practice would but unnecessarily fatigue the patient.

When the cavities are dry and of fixed size, inhalations should not, in my opinion, be practised. Alcohol, whose power in producing fibrous tissue is in such cases of great value, is specially beneficial, and calcic phosphate should also be now given in large doses, specially in combination with milk. If, however, the digestion is in any way affected, or the constitutional treatment of which alcohol, glycerine, and creasote are the principal agents is in any way compromised, I do not continue to administer the latter remedy.

The process of excavation is at times so acute, and the pyrexia so nearly continuous, that the patient must necessarily remain in bed; but when once improvement shows itself, the previous habits should be recommenced as soon as possible, and the moments of apyrexia employed for giving to the patient the benefit of out-of-door life at a suitable residence, the exercise taken being proportional to the strength of the person, and continued as long as this permits. Such is a general principle which may be applied to every period of the complaint, and the time of confinement to the house should be deferred as long a time as possible, and when this moment arrives the day must not be spent in the bedroom, which during this time is to be thoroughly ventilated.

Nor is this all, since when the time of ulceration commences another measure should be adopted, equally useful to the patient and to those who are with him. On the evening of each day an antiseptic atmosphere is produced in the bedroom by means of spray from a ten per cent. solution of carbolic acid, as much water being used as would fill once or twice, according to the size of the room, the bottle belonging to the apparatus used. This practice is useless in the case of patients who inhale, but in other cases may be really looked upon as constituting a form of local treatment, since during the whole night respiration is carried on in an atmosphere filled with particles which are medicinally beneficial. It is but rarely, according to my experience, that the odour of the atmosphere has a bad effect upon the patient. If, however, this happens, inhalation need not be abandoned, but benzoate of soda should be used in the place of carbolic acid. This salt has no odour, and its antiseptic properties have been well shown to exist by Klebs; at the same time, in this special case it is less beneficial than carbolic acid, since on account of their weight the saline particles are almost at once precipitated instead of remaining diffused through the surrounding air.

It will thus be understood that the time during which caverns form should not be one of therapeutic inaction. This period, like others, gives indications which should be fulfilled, and to do this is the strict duty of the medical attendant, whatever may be for other reasons the final result of his efforts.

Such are the general rules which guide me in the different periods of pulmonary phthisis, representing together what I would willingly call, but for the pretentious meaning of the word, the ideal of my therapeutic treatment; too often, however, as much in the apyretic as in the febrile period, the pursuit of that ideal must be temporarily abandoned on account of certain pathological incidents connected with the development of tubercles. The most important of these events, which are real complications affecting both the treatment of the disease and the patient, are dyspeptic troubles and hæmoptysis, and the treatment which, in my opinion, should be adopted in such cases, will now be briefly considered.

As regards disorders of the digestion, I should say that these are comparatively rare in patients living in the country, who observe rigorously the hygienic precepts already laid down. In town patients, however, they are, on the contrary, most frequent, and often prevent the fundamental plan of treatment being carried out, to the great detriment of the patient. Loss of appetite and dyspepsia due to some digestive disorder are the most common occurrences, imposing the temporary abandonment of remedies, and often even of milk. The mistake must not be made of considering this condition to be always a simple form of functional dyspepsia. Such an erroneous idea would lead to an accumulation of so-called peptic (*eupeptiques*) means of treatment being adopted, which inconvenience the patient and in most cases produce no result, from the fact that dyspepsia, supposed to be functional, was in reality a symptom of gastric catarrh. If this catarrhal disorder of the stomach is treated, success rapidly follows.

When the condition presents itself as an acute disorder, and the tuberculous affection, still at its commencement, has not as yet compromised the patient's strength, an emetic, such as ipecacuanha alone, or combined with tartar emetic, should be employed, that any irritation of the intestine may be avoided. If this catarrhal condition is subacute, or the employment of an emetic seems to be counter-indicated, aperient remedies should be administered more or less repeatedly according to the requirements of the case, and always in small doses. When the indications of catarrh, specially the white coating of the tongue and bitter taste in the mouth, are pronounced, the existence of slight diarrhœa is no obstacle to the treatment, merely indicating, on the contrary, that catarrh of the intestine co-exists, which purgative remedies would be the best means of removing. Castor oil or drastic medicinal agents should on no account be administered, but effervescing salts or the natural mineral waters should be employed, half or three-quarters of a glass being taken on the morning of each or alternate days according to the effect produced. When the indication is precise the appetite will probably return in less than a week, and the digestion will have so improved that this treatment can be gradually resumed. Such a form of catarrh exists in most cases of the dyspepsia which attends phthisis, coming on at any period of the complaint, and to be then opposed according to the above principles, unless the disease be now at an advanced stage, or there are positive signs that the intestine is ulcerated.

Cases, however, exist in which the dyspeptic symptoms cannot be referred to gastric catarrh. Dyspepsia would then exist, understood in the usual sense of the word, since the altered condition of the glands and secretions, the change in the excitability of the muscular fibres which are the true cause of the so-called essential disorder, cannot be directly appreciated, only being supposed to exist by a somewhat vague induction based upon the physiological analysis of the symptoms. No fixed rule can be applied to cases of this kind, the same uncertainty, variableness, and doubt which exists in ordinary dyspepsia being found to exist here also. Without leaving the subject now discussed, the multiple indications of remedies which have opposite effects cannot be considered. The febrile symptoms and those due to absorption of putrid matter being understood, the existence of tuberculosis would not affect the treatment; dyspepsia exists and must be treated in the ordinary way. Thus, according to the case, absorbents, alkalies, acids (hydrochloric and lactic), stimulants (strychnia, nux vomica), sedatives (opiates, potassium bromide in small doses), purgative salts, not in an aperient but in a laxative dose, or lastly, preparations of pepsin should be administered. Cod-liver oil would not naturally be given to a tuberculous patient who suffers from dyspepsia, but should no gastralgia exist, the continuous employment of creasote may be tried. There is no counter-indication to its use except the existence of pain, and if this be absent the remedy may be well borne for some time; at all events, the experiment should be tried. If the patient has been previously undergoing hydrotherapeutic treatment, this must on no account be abandoned, since it may help to re-establish the functions of the stomach. With regard to food, as in the case of medicinal treatment, no fixed rule can be made. In my opinion, the attempt should be first made to continue the ordinary diet of the patient; but when after some time, and notwithstanding appropriate treatment, but little improvement takes place, some uniform diet should be employed. The same variety exists with regard to this, and in some cases the diet should consist exclusively of milk, in some of raw meat, while in others no choice can be made until after an observation of some days. As in all serious forms of dyspepsia, the best diet is that which is most easily borne, and often it cannot be previously determined what this would be. In obstinate cases, counter-irritation over the region of the stomach may be most beneficial, and should always be practised, either in the form of heated metallic points or blistering agents.

It is an important fact, however, that this obstinate form of dyspepsia, which so rapidly aggravates the condition of the patient by affecting the nutrition and interfering with the treatment, is scarcely observed except in the confined air of towns, and in many cases residence in the country will alone cure symptoms which have resisted the most appropriate treatment. Dyspepsia may be also found to continue in patients who have had the misfortune to reside where the climatic conditions are such as to provoke and keep up disorder of the digestive organs. In such a case it is a mistake to unreasonably continue treatment which cannot but fail, and immediate change of residence is the only measure which will be really effective.

In the advanced stages of the disease, food cannot for many reasons be given in the ordinary form. Milk is rarely borne, or at any rate not in sufficient quantity to satisfy the requirements of nutrition; and raw meat, to which a sharp taste is given by means of salt, alcohol, or sugar, according to the case, supplies the last resource. In many cases nourishment can be thus effected at least in a degree proportional to the organic expenditure, which is always but slight at such times; and not rarely it is possible, owing to this diet, which must be looked upon as medical treatment, to return to more varied and attractive food. If this means fails, recourse may now be had to peptone, made artificially from meat, and of which the nutritious properties have been clearly established by numerous observations made in foreign countries, as well as by researches made in France, and notably by Daremberg and Catillon. It is also certain that peptone does not lose its nutritious value when administered by enema and not by the mouth, and thus supplies a valuable resource when food cannot be taken in the ordinary way. Even without considering such extreme cases, peptone may be useful in the various dyspeptic conditions which have just been considered, being a food which is already suitable for absorption, and which may fully compensate for that forced deficiency which exists in the ordinary diet.

Vomiting, usually due to gastric disorder, presents no special indications, the treatment being that of the dyspepsia to which it is due. It may be said, however, that counter-irritants are now of special value, and that blisters, alone or combined with morphia, rapidly remove this troublesome symptom, though it may already have resisted persevering medical treatment. When the vomiting immediately follows the ingestion of food, denoting irritability of the stomach, the effect of counter-irritants may be increased by administering two or three minims of laudanum five minutes before meals, a practice which I have often found to succeed. When vomiting takes place without any obvious dyspepsia, resembling the so-called nervous vomiting, ether spray, applied twice or thrice daily for six minutes at a time, three over the region of the stomach, and three over the corresponding part of the spinal column, may prove more beneficial. Without looking upon this means as a substitute for all others which might be employed in the same circumstances, I expressly recommend it from having found it to succeed when every other mode of treatment has failed.

That form of vomiting which occurs even more frequently, and which is mechanically produced by the movements of coughing, is unaffected by remedies which act directly upon it. To relieve the cough is the only process which makes it cease. Opium, belladonna, potassium bromide, chloral, are the most useful agents, while subcutaneous injections of morphia should be avoided, and if their use becomes necessary should on no account be employed as a daily remedy. The habit of employing morphia in this way renders a continual increase of the amount injected necessary, and this in time produces the condition of morphism with all its consequences

Amongst the means of treating cough the employment of counter-irritants directed against the local lesion which produces it should not be forgotten.

Diarrhœa is often connected with gastric dyspepsia, and then requires no special treatment. In other cases it is due to intestinal catarrh, and if the condition of the patient permits, should first be opposed by saline purgatives in small doses. Should this treatment be impossible or fail, absorbent powders, bismuth, or laudanum should be employed. The two latter will be taken by the mouth, or in an injection according to the case, and the presumed seat of the intestinal catarrh. Should the complication persist, a uniform diet composed of broiled or raw meat should be instituted, milk being rarely well borne, unless it be that of the ass. In the same conditions tincture of iodine applied over a large surface of the abdominal wall may be found useful, or in the most persistent cases the application of collodion. Their application is certainly inconvenient, but appears to be of real benefit. During the existence of these complications the fundamental plan of treatment must necessarily be abandoned, creasote being alone excepted in combination with glycerine, as has been already explained, since in the absence of gastric dyspepsia this remedy is found to have no effect upon the diarrhœa which exists. Another form, which is more rare, seems due to incomplete digestion of the food, the fetid condition of the intestinal contents, indicating fermentation of a putrid character, and characterizing the diarrhœa which exists. A few doses of salicylic acid form the best kind of treatment. The catarrhal state which is necessarily associated with this disorder may persist when the diarrhœa has ceased, but will be quite amenable to the ordinary means of treatment. When the diarrhœa depends upon intestinal ulcers, it may be improved temporarily, but will quite baffle every effort to cure it.

A few lines will now be written upon the treatment of hæmoptysis. This should not be the same in all cases, and some fundamental distinctions are absolutely necessary. It is specially requisite to separate the hæmoptysis which occurs before the formation of caverns from that which takes place later in the disease. The former is due to congestion, while the latter most often follows erosion or rupture of the vessel. In the former, which will be first considered, two groups must be considered, according as the hæmoptysis is attended by pyrexia or not. Certain measures which may be employed in all cases being excepted, such as the half-recumbent position of the patient in bed, immobility and silence, the ingestion of ice and iced drinks, lemonade, water of Rabel (*eau de Rabel* *), the maintenance of a cool temperature in the room, and the proscription of pillows or mattresses which contain feathers, the treatment should vary according to the different conditions which I have named.

The hæmoptysis which is unaccompanied by fever is of slight intensity, and usually of short duration. In most cases it suffices to add to the preceding means of treatment a draught containing from half a drachm to a drachm of the liquid extract of rhatany, or from twenty to thirty drops of solution of ferric perchloride, or, again, from fifteen to thirty grains of gallic acid, in order to arrest the hæmoptysis. If, however, the hæmorrhage has a more serious character, and though unaccompanied by fever, persists from thirty-six to forty-eight hours after the adoption of this treatment, more should, in my opinion, be done. The chest should be covered with a large blister, and dry cupping should be practised, cupping-glasses being applied at least twice daily to the number of from forty to sixty over the lower extremities, and at the base of the chest. Opium should also be administered in large doses from four to six

---

* The "eau de Rabel" is composed of alcohol (three parts) and sulphuric acid (one part). From five to thirty drops are taken in a mucilaginous drink.

grains (20 à 40 *centigrammes*), being divided into pills of which each contains one-third of a grain (2 *centigrammes*). One of these should be taken hourly until a tendency to sleep is produced. On the following day the dose is altered in proportion to the result obtained and the narcotic effect, and thenceforth the amount is gradually diminished, but in such a way that the remedy is not completely omitted until at least three complete days have passed without the occurrence of hæmorrhage. If hæmoptysis occurs unattended by fever before the time at which caverns form, and in sufficient amount to cause real anxiety, the patient must not be only subjected to treatment which is so slow in producing its effect. Counter-irritants should be employed at the same time, while the subcutaneous injection of ergotine, and inhalation of perchloride of iron, of which from forty-five minims to a drachm are dissolved in six ounces of water, would be also beneficial. Such an event must be most rare, since I have never seen it occur, and I only mention it in order that the effect of the above-mentioned treatment may not be exaggerated.

There is another mode in which hæmoptysis may occur, without the existence of pyrexia, as not unfrequently takes place, and to which attention should be given. Unattended by fever for one or perhaps two days, on the third, notwithstanding the treatment adopted, pyrexia occurs, and the hæmoptysis continues, becoming perhaps even more intense. In this case I at once relinquish the original treatment, and employ the same remedies as in hæmoptysis which from the first was associated with pyrexia.

Usually in large amount, the hæmoptysis is in many cases repeated, and always constitutes a pathological complication of several days' duration; theoretically, it furnishes on all occasions at the onset an indication to bleed, on account of the pulmonary congestion which exists. Removal of blood lessens the congestion, diminishes the intravascular pressure, and accordingly checks the flow of blood, and prevents any fresh rupture from taking place. In practice, however, this means of treatment should only be employed in those who are still of strong constitution, and who are not already weakened by tuberculosis of long duration. Even in that case it should be reserved for those cases in which the large amount of blood lost indicates the necessity of active treatment with immediate effect. Whether the patient be bled or not, which in any case would be but the first part of the treatment, blisters should be applied successively to different parts of the chest wall, and dry cupping be practised over the same parts. Ipecacuanha should then be administered in the following manner: a grain and a half (10 *centigrammes*) are to be taken every quarter of an hour until nausea is produced without vomiting. When this result is obtained the remedy should be taken less frequently, as every half-hour, hour, or even two hours, according to the state of the pulse and temperature, and the tendency to vomit. When reduction in the temperature and diminution of the hæmorrhage are sufficiently pronounced to constitute a decided improvement, the remedy is no longer employed, to be recommenced, after some hours of rest, should the return of pyrexia and the condition of the pulse indicate a tendency to hæmorrhage. At the same time the strength should be sustained by means of iced wine and cold soup, a method which I have found most successful, notably in the case of the Russian prince residing at Auteuil, who has already been mentioned.

It should be known that the employment of this remedy is not without danger, specially in the case of those who are affected by phthisis at an advanced stage. In a very short time the reduction of the temperature may be replaced by true collapse, and thus the effects of the treatment should be carefully watched, with the thermometer in hand, the condition of the heart being also taken into serious consideration, as well as the tendency to perspiration which exists.

Other nauseating remedies, such as tartarated antimony in small doses, or kermes * (*kermès*) to the amount of from four to six grains daily, produce analogous effects, and may consequently be no less beneficial. At the same time, it has always seemed to me that the effect of ipecacuanha is more rapid and decided, besides which it is not equally liable to produce diarrhœa, which preparations of antimony are always prone to do even in the smallest doses.

Notwithstanding the good effects of the remedies mentioned, they should not, in my opinion, be employed in the following cases: firstly, when vomiting occurs after the second or third dose, and persists though but half the amount, namely, gr. ¾ (5 *centigrammes*) of ipecacuanha, is given; secondly, when the patient is so weak that some danger exists that the condition of nausea may become one of collapse; thirdly, when the special conditions of the patient prevent that careful attention which is necessary; and lastly, when the hæmoptysis, although attended by pyrexia, and presenting in consequence the acute character of hæmorrhages belonging to this class, is not so abundant and intense as to require treatment which acts with special rapidity. Since nauseating remedies are extremely unpleasant, it is in my opinion more to the interest of the patient to refrain from them, their necessity not having as yet been established.

In these different circumstances I continue the same treatment as regards counter-irritants, iced drinks, etc., but administer in the place of nauseating remedies quinine sulphate, or ergot of rye. According to the intensity of the fever, from fifteen to twenty grains of the former remedy should be administered on the first day within the space of thirty or forty-five minutes, in order that the whole effect of the dose may be obtained as already explained. A smaller dose should be given on the second day, and a smaller still on the third should

* See note, p. 188.

no change occur, the temperature being taken as guide with respect to the advantage of returning temporarily to the same large doses as at first. If the sulphate is not well borne; the hydric bromate should be employed, and each dose should then be increased by seven grains as compared with the sulphate. Should the remedy be so ill tolerated that it cannot be used, subcutaneous injections must be employed according to the rules already stated. Quinine or ergot of rye will be preferred according to the respective importance of the pyrexia or hæmorrhage. If the temperature reaches from 102° F. (39° C.) to 104° F. (40° C.) while the hæmoptysis is of slight amount quinine should be given, whilst if the fever is slight and the hæmoptysis more or less profuse, ergot of rye should be administered to the amount of from four to eight grains every two or three hours until some good effect is obtained, or tingling and numbness are produced in the fingers and toes.

Lastly, when the hæmorrhage belonging, as shown by the fever and other characters which it presents, to the group which we have just considered is profuse enough to cause immediate danger, the nauseating and antipyretic remedies should both, in my opinion, be omitted, and subcutaneous injections of ergotine should be combined with inhalation of perchloride of iron. Digitalis should never be administered in the treatment of hæmoptysis occasioned by tubercles. While it doubtless reduces the temperature and diminishes the rapidity of the pulse, and hence may be seemingly compared with other antipyretic agents, on the other hand the heart's contractile power is increased, so that while diminishing the fever it may favour the continuance or return of the hæmorrhage.

The hæmoptysis which occurs at the later period when caverns form, by whatever symptoms it may be accompanied, requires the employment of the most powerful hæmostatic agents, being usually due to the rupture of an aneurism or diseased branch of the pulmonary artery. The hæmorrhage is a cause of immediate danger, all else being of secondary importance, and dry cupping largely employed (the cupping-glasses of Junod [*les ventouses de Junod*] being used, if they can be easily obtained and the patient seems able to bear their application), the injection of ergotine, and inhalation of perchloride of iron should be the treatment adopted. If diminution of the hæmorrhage is not rapidly obtained, the chest should be covered with ice on account of the imminent danger, and though objecting to this practice, which has been proposed in the treatment of ordinary hæmoptysis, in the above special conditions I look upon it as legitimate and useful. At the same time, it must be considered an extreme measure, since the form of hæmorrhage which is due to rupture of a vessel within a cavern is almost constantly fatal, if not upon its first, upon its second occurrence, which usually follows within a brief period. It is even possible that this form of hæmorrhage may prove fatal before any treatment can be adopted.

CHAPTER IX.

TREATMENT OF THE ORDINARY FORM OF PHTHISIS (*Termination*).

TREATMENT OF PNEUMONIC PHTHISIS; OF ACUTE MILIARY TUBERCULOSIS.

Inhalation of benzoate of soda in ordinary phthisis—Researches and personal observations—Modes of application—Practical difficulties—Effects upon the general health and local lesions—Conclusion.
Pneumonic tuberculosis or phthisis—Its distinctive characters—Acute onset of disease—Chronic period—Varieties in the onset and evolution.
Principles of treatment—Fundamental indications—Stimulating remedies—Antipyretic remedies—Mode and processes of administration—Special indication of digitalis—Counter-irritant treatment.
Acute miliary tuberculosis—Possibility of cure—Observations of authors—Personal observation—Principles and means of treatment.

THE treatment which should be employed in the different stages of ordinary phthisis has been explained in all its details, and before passing further I would communicate the results of my experience with respect to the new plan of treatment recommended in 1879 by Rokitansky, of Innspruck, and his assistant Dr. Kroczak, which is founded theoretically upon the supposed connection between bacteria and tubercle, practically upon the inhalation of benzoate of soda, the destructive effect of which upon bacteria has been established by Klebs and Schüller. Upon becoming acquainted with this mode of treatment, I at once determined that my patients should profit by it, and in order that it might be employed with the requisite precision, I requested Professor Rokitansky to make me acquainted with the details of its administration. The request was at once granted, and I am pleased to have this opportunity of publicly thanking him for the favour.

The quantity of benzoate inhaled daily should be fifteen grains (*un gramme*) for each thirty-five ounces (*par kilogramme*) in the weight of the body, that is to say about two ounces (50 *grammes*) in most cases. The solution of the salt should be in the proportion of five per cent., so that the two ounces (50 *grammes*) would be dissolved in forty ounces (1000 *grammes*) of water. The inhalation is made in two sittings, one in the morning and the other in the afternoon, the liquid in each case containing an ounce (25 *grammes*) of the benzoate dissolved in twenty ounces (500 *grammes*) of water. When this treatment is adopted, the apparatus which I habitually use would not be employed, since at least two hours and a half would be required to turn such a large quantity of liquid into vapour, five hours being thus spent daily by the patient in the practice of inhalation. Having ascertained these facts, I did not think it right to recommend a process which is so troublesome, not to say dangerous. Ascertaining how much vapour was added to the solution employed whilst the apparatus was in use, I therefore commenced by employing a ten per cent. solution, an ounce (25 *grammes*) of the salt being dissolved in ten ounces (250 *grammes*) of water, so that, considering the addition of the vapour, the liquid inhaled was a five per cent. solution. In these conditions each sitting lasts in the mean an hour and a half; but though I have found it possible to employ this mode of treatment in the case of two or three persons who were exceptionally patient, I soon discovered that it would be impossible to continue these observations unless I simplified still more the treatment employed. The patients invariably refused to submit to it, and although those that were most reasonable adopted it for one day, even these manifested after that time an insuperable repugnance to such an inconvenient practice. This being the case, I attempted to do more. After some trials I found that an ounce (25 *grammes*) of the benzoate could be dissolved in six ounces (150 *grammes*) of water without the formation of any precipitate, that the time required to convert this liquid into vapour, including the interruptions due to cough and expectoration, was about three-quarters of an hour, and I therefore adopted the following plan. The quantity of water added while the apparatus is in use being naturally equal to the loss on the part of the solution, an ounce (25 *grammes*) of the salt was dissolved in six ounces (150 *grammes*) of water, so that a twelve per cent. solution was inhaled. I certainly regret having to modify the plan employed by my learned colleague of Innspruck, but there was no choice in the matter; I met with difficulties which he does not seem to have experienced, so that I was obliged to adopt the plan mentioned above or to abandon the treatment. The patients, therefore, under my care inhaled in the mean one ounce (25 *grammes*) of benzoate dissolved in six ounces (150 *grammes*) of water.

In all other details I followed strictly the plan of Rokitansky. The production of spray was arrested whenever an attack of coughing supervened, in order that the loss might be reduced as much as possible, and when the patient had ceased to inhale the vapour, he still remained for at least an hour in the same atmosphere, so as to continue breathing during that time an air charged with saline particles. While this treatment is being carried on, Rokitansky causes the patient to be supported by good nourishment, milk, cod-liver oil, etc. It seemed to me that I should better appreciate the effects of the benzoate by employing it alone, and so long as I could do so with justice to my patients I abstained from any other form of treatment, except the diet and wine.

I thus treated fifteen patients who were affected by tuberculosis in different degrees; in the greater number softening was either beginning to occur or had occurred for some time, caverns existing in four cases. Of the latter, three were unable to tolerate the inhalations for more than three days on account of their debility; two of the twelve remaining cases could only do so for a week, on account of the inconvenience attending the practice; while the ten others continued the treatment, one for fifteen days, four for a month, four for two months, one for three months and a half.

The results of these trials, I regret to say, did not answer my expectations. The expectoration I certainly found to diminish, and become of better quality, as Rokitansky had done, and, except in the case of the patient who was only treated in this way for fifteen days, an improvement in the general condition took place, as shown by the appetite, the integrity of the digestive functions, the external appearance, and three times by increase in weight. This result, however, I cannot attribute entirely to the benzoate, since in patients of this class the life and diet at the hospital, which is far better than in their ordinary conditions of life, may alone effect this improvement when the disease is not in too advanced a stage. Again, when my three first trials had shown that this new plan of treatment had no special effect, I did not feel justified in abstaining from the employment of other therapeutic measures, such as cod-liver oil or glycerine, and alcohol. This being the case, it seems to me unwise to form a definite conclusion in favour of the benzoate, the only ascertained fact in connection with which is, in my opinion, the modified expectoration. It would certainly act to some extent as an antiseptic; but the result in this respect did not seem more pronounced than when carbolic acid was inhaled, a practice far less fatiguing to the patient on account of its shorter duration, or again in many cases when creasote was perseveringly administered.

My hope that the inhalations would, after some length of

time, have an antipyretic effect was also disappointed. In patients with the disease but slightly advanced, in whom such a slight elevation of the evening temperature as to 100·5° or 100·8° F. (38° or 38·2° C.) had taken place, the pyrexia undoubtedly ceased. Might not this, however, have been due to rest ? an explanation which seems the more true, since when the pyrexia was intense no effect was obtained, so that recourse to quinine and salicylic acid was often necessary.

As respects the local lesion, I feel able to speak positively, since in none of the patients could any perceptible improvement be recognized, while in many the extent of the lesions was found to show some increase. In one patient alone in whom the disease was least advanced, after a month's treatment the following modification occurred. Catarrh existed at the two apices; upon his admission crepitation was heard during ordinary respiration without the movement produced by cough being necessary. At the end of a month it was only heard when the patient coughed, ordinary breathing being unaccompanied by any morbid sound. This difference, however, was only one of degree (the patient now leaving the hospital), and not more than would be expected to occur in this time from the effect of suitable treatment aided by rest and nourishing diet.

In two patients hæmoptysis took place during the course of the treatment. One had been permitted to go out on the previous evening, and, being a musician by profession, had played the flute during many hours, which without doubt was the true cause of this complication; the other, however, in whom the tubercles were beginning to soften, had remained quiet at the hospital, without exposure to any disturbing influence. Hæmoptysis now occurred for the first time, and though I would on no account form any conclusion from this single fact, I thought it right to mention its occurrence.

To sum up, the treatment is far from being always accepted,

and when employed is undoubtedly a cause of fatigue. Nor, according to my experience, do the results obtained compensate for the difficulties encountered. Its utility in cleansing the bronchio-pulmonary surfaces appears to be certain, but it seems also that this result may be more easily obtained by means of treatment by inhalation, which is less prolonged.

Inhalation is not itself disagreeable, either on account of the taste or impression which it produces upon the mucous membrane, and my own personal experience has often convinced me that benzoate of soda is, in taste at least, superior to carbolic acid. Since much smaller doses of the salt are required for its cleansing action upon the mucous membrane than for its antiseptic effect, this salt may be beneficially employed by patients who, though they require such treatment, cannot tolerate the acid. It is sufficient in these cases to have from five to six drachms (20 *ou* 25 *grammes*) inhaled in three or four sittings during the course of the day.

This subject will be concluded by the account of a fact which is interesting from different points of view. In three patients I assured myself that the benzoate was truly absorbed, inasmuch as I found more than fifteen or twenty grains (*une quantité superieure à* 1 *gramme et* 1 *gramme et demi*) of hippuric acid passed during the twenty-four hours, the normal quantity passed during that time being from four to six grains (*de* 30 *à* 40 *centigrammes*). This increase is due to the fact that in these circumstances benzoic is changed into hippuric acid.

Pneumonic Phthisis.—The peculiarities which distinguish this disease will be briefly recalled before the question of its treatment is discussed. It is essentially characterized by its onset, which is sudden, acute, and presents the same symptoms as unilateral acute pneumonia, whether circumscribed or disseminated, lobular or lobar. Resolution, however, does not occur at the ordinary time, the fever, the symptoms of acute illness, the signs given when auscultation and percussion are practised persisting with slight variations for a month, six weeks, or even more; and the pneumonia, believed at first to be acute, thus shows itself to be a tuberculous or caseous form of the disease. When this condition has lasted for a certain time (some days or weeks) beyond the maximum of time occupied in the resolution of ordinary pneumonia, it follows one of the four following courses: 1. Recovery occurs, gradual improvement taking place at a late date, with or without remaining indications of disease as recognized by means of the stethoscope. 2. It persists as an acute illness, and compact infiltrations of the lung occur (lobar or lobular hepatization), softening, and excavation. The patient succumbs in a few months, after passing in a relatively short space of time through the same stages as occupy several years in their evolution when chronic or ordinary phthisis exists. It is for this variety that the name of galloping phthisis should be reserved. 3. In other cases the acute illness ceases after a longer or shorter time, usually from six weeks to three months, the disease then assuming all the characters of chronic phthisis. 4. Lastly, the pneumonia proves at times fatal during the acute onset in a period which varies from fifteen to forty or sixty days. In all these cases the onset is possibly marked by attacks of such abundant and repeated hæmoptysis that a hæmorrhagic form of the disease may be admitted to exist.

Exceptionally—and this is a fact which it is well to dwell upon, since it is less known—the sudden and acute onset which usually characterizes the disease is absent. The onset is perhaps attended by pyrexia, but the fever is less intense, the accession of symptoms less complete, their development occurs more gradually, and the patient is not at the end of twenty-four or forty-eight hours in the same serious condition as in ordinary acute pneumonia. The character of the case resembles rather one of ordinary broncho-pneumonia, and an erroneous diagnosis can only be avoided by the consideration that in this disease the affection is unilateral. Whatever, in fact, may subsequently occur, the lesions in pneumonic tuberculosis are at first limited to one side, and according to my experience more often to the right than to the left.

Another event of still more uncommon occurrence, which has not, as far as I know, been yet pointed out, but which I have certainly twice observed, is the absence of any intense fever during the first stage of the disease. The onset occurs gradually as in ordinary phthisis, without the patient taking to his bed unless ordered to do so, but a feeling of discomfort is felt, of fatigue with wandering pains on one side of the chest, and often slight feverishness in the evening or at night. He begins to cough, and when the chest is examined some weeks after the commencement of the illness there is no indication of catarrh at the apex, or of discrete and disseminated miliary tubercles, but compact and homogeneous infiltration of variable extent in one of the lungs. Such is the characteristic lesion of pneumonic phthisis with slow and quiet onset. The chronic onset of the disease, which is quite exceptional, does not take away the liability to acute pneumonia and broncho-pneumonia, which occur so frequently in the course of this form of phthisis. The patient is subject to the same complications, and this uncommon variety of the disease only differs in reality from the ordinary form by its mode of onset. These facts should be known, that pneumonic phthisis with gradual onset may not be mistaken for the granular or ordinary form of the disease.

Without then considering the mode of onset, but according to the subsequent progress of the disease, two forms of pneumonic phthisis may be distinguished ; a rapid form in which recovery seldom occurs, and which, when it ends fatally, may do so before or after ulceration of the lung, and a slow form which, after an acute period of long or short duration, has a chronic course resembling that of ordinary phthisis

The therapeutic treatment is identically the same in the two diseases, and when once this form of pneumonic phthisis has become chronic, and the acute symptoms of the onset have ceased, the condition of the patient is the same as when the disease, chronic from the first, has reached that stage at which the same local lesions exist. The pneumonic form of phthisis furnishes, as I have said, special indications for treatment by mineral waters and climate, as will soon be repeated. So far, however, as the hygienic and medicinal treatment are concerned, the conditions are the same in the two cases, and after a somewhat stormy outbreak of the disease, varying in duration, the phthisis becomes chronic, and should be treated accordingly. This subject will not be reconsidered.

With regard to the acute onset of this form of phthisis, one of two results, unequal in importance, should be aimed at: firstly, complete resolution at any moment of the acute onset, namely, recovery; secondly, the passage of the disease into a chronic condition in which the imminent danger of the acute condition no longer exists, and the prognosis is relatively more favourable, enabling the patient to derive benefit from every beneficial event which occurs in the disease. Though this result is less favourable than the preceding, it is also specially desirable in this disease, since observation has led me to believe that both relative and absolute recovery are more possible in the chronic stage of this form of pulmonary phthisis, than in miliary tuberculosis which is chronic from the first. The latter, though of far slower progress, is more undoubtedly progressive, while the former, though often fatal at the time of its onset, may on the other hand become a chronic affection, and, remaining inactive and torpid, present more tendency to the stationary condition, with even partial repair of the existing lesions. This, in fact, is what my observations have brought me to believe. The repair may take place after a long time, but while the stationary condition persists it is always possible, and this chronic period in the course of pneumonic tuberculosis is therefore of great importance.

With regard to recovery, more or less deferred during the period of acute onset, namely, for two, three, or four months, this termination, though of much more uncommon occurrence, is also, in my opinion, possible. Though I have seen no such event during the last few years, cases are reported in my Clinical Observations, and in these, as I have already had occasion to say, on account of the age of the patients, their previous history, the course of the symptoms and the character of the lesions, it seems impossible that acute broncho-pneumonia of long duration can have been mistaken for tuberculous pneumonia.

The results, then, which should be sought during the acute onset of pneumonic tuberculosis can really be obtained; one however, is exceptional, and its existence perhaps in some cases impossible, while the other, although uncommon, is nevertheless of relative frequence, and this fact should at any rate encourage the physician.

I treat the acute onset of pneumonic tuberculosis in the same way as a severe attack of ordinary pneumonia. During the first few days, as is natural except in those rare cases in which tuberculous pneumonia assumes a hæmorrhagic form, it is impossible to distinguish whether the case is one of simple pneumonia or broncho-pneumonia, or the same complaints of a specific kind. Whatever, however, the duration of the acute period may be, I continue the same treatment, increasing the strength of the remedies employed in proportion as the disease exceeds by a longer period the time within which resolution occurs in ordinary inflammation of the lung. The treatment has certainly no effect upon the tuberculous affection, the hope being that the patient may be enabled to resist the acute attack, and reach the period at which improvement will take place, should it do so at all. This fortunate event does not occur in the galloping form of the disease, which proves fatal to the patient before the acute onset has really terminated. Such being the only object which can be pursued, my first consideration is the condition of strength; and since debility is a more or less early though certain effect of the disease, I do not await its occurrence before opposing it, but make an effort to prevent or diminish it by means of stimulating remedies, when once the absence of recovery at the usual period gives rise to some suspicion as to the nature of the affection. Pneumonia may undoubtedly exist, which is not tuberculous, and in which resolution occurs at a late date; but even then this treatment is, in my opinion, the most advantageous, and by its early employment the patient receives the benefit of appropriate remedies at a time when they are rationally indicated, and the mistake of waiting until signs of debility show themselves before employing agents which might avert them is thus avoided. In many cases the indication of this treatment shows itself at a still earlier period, pneumonia, the nature of which could not be ascertained until a somewhat later date, occurring suddenly in an adynamic form, or with serious symptoms which undoubtedly indicate the existence of debility. When this is the case I employ stimulants from the time that the patient is first seen, and without waiting until more prolonged observation reveals the true nature of the complaint. The same rule directs my treatment in the case of typhoid fever. To act powerfully does not suffice, but it is necessary to act at an early period of the disease.

Having said this as to the principles of treatment, the plan which I adopt is the same as that explained at much length in my Clinical Lectures, as far as its fundamental object is concerned. Observation, however, has led me to make an addition thereto which I consider most important, namely, that of antipyretic agents, so that stimulating and antipyretic remedies, employed from the first and continued without cessation, constitute the treatment which I now employ.

I exclude every means of treatment which might weaken the patient, not even permitting broth to be taken, which completely removes the appetite, and aggravates the gastric catarrh which necessarily accompanies every condition of fever. As drink, I administer red wine combined with water in the proportion of at least one to three. Whatever the intensity and type of the fever may be, the patient should be nourished, in my opinion, according to the degree of inappetence and intensity of the gastric catarrh, by milk, beef-tea, jelly or gravy made from meat, and wine of Bordeaux in varying quantity. Well-prepared meat-jelly, without isinglass or gelatine, is also of great value. It nourishes without causing fatigue, and if combined with orange or lemon juice is pleasant to the taste, leaving in the mouth a sensation of freshness which diminishes for some moments the feeling of heat produced both by the pyrexia and remedy. This diet is usually accepted without difficulty, but since it must be continued for a long time, the patient is often seized with insuperable repugnance to any food which is made from meat, in whatever form it may be given. Milk then becomes a valuable resource, and by drinking it in small quantities the patient will succeed in taking within the twenty-four hours all that is required for the purpose of nutrition. In many diseases I recommend this nourishment from the first, but in that which we are now considering I prefer the jelly and soup made from meat as long as they can be taken in sufficient quantity. When recourse must necessarily be had to milk, I usually administer wine without the addition of water, in the form of sherry, Madeira, port, or Malaga in the place of claret, as I also do when the latter wine has become irksome to the patient.

It is useless to insist upon the necessity of cleanliness, proper ventilation of the room, an antiseptic atmosphere, frequent renewal of the bedclothes, and attention to the digestive functions, since there is nothing special to be said upon these points. With regard to remedies, these are usually three in number—alcohol and cinchona bark, which are administered permanently, and quinine or salicylic acid at intervals according to the temperature of the patient. Exceptionally, and in order to counteract a particular danger to be soon specified, digitalis is employed as an antipyretic agent.

When it is foreseen that the treatment must be continued for a long time without notable change, the mode of its employment is of great importance on account of the more or less rapid distaste which the patients feel, and which produces a serious obstacle to the administration of remedies. For many years it has been my plan to adopt the following process, which experience still shows me to excel any other. Instead of administering alcohol, and cinchona bark in a mucilaginous preparation, which is itself without effect, difficult to digest, and likely before long to be ill tolerated, I use as vehicle a mixture analogous to the cordial draught (*potion cordiale*) of the Paris hospitals, in which I replace common syrup by syrup of orange-peel. The composition is as follows: Old red wine, four ounces; tincture of cannella alba, from half a drachm to a drachm and a half according to the taste of the patient; syrup of orange-peel, from an ounce to an ounce and a half. To this vehicle, itself an active agent, alcohol is added in the form of old brandy or rum, at a dose which varies from one to three ounces according to the special conditions of each case, that is to say, the sex, age, and habits of the patient, and more than all the degree of pyrexia and strength. When as much as two ounces of alcohol are given, the proportion of syrup should be increased so as to diminish its impression upon the mucous membrane of the mouth and pharynx.

To this vinous drink, which is mixed with spirit, I add in all cases from thirty grains to a drachm of the soft extract of cinchona (*de l'extrait mou de quinquina*), and this mixture is given in doses of a table-spoonful (*par cuillerées à bouche*) every two hours. Since it contains nothing which can interfere with digestion, there is no need for consideration at what time meals should be taken, and sleep alone should interrupt the employment of the medicine. Thus not only is the effect of this remedy powerful, but the ease with which it can be employed is of special value in all cases, and specially in those of long duration.

The treatment being once instituted, is continued until the end of the disease. Whatever complication may arise, no departure is made from a plan which, in my opinion, alone offers any chance of success. Though the cough may increase from the effect of alcohol, which not rarely happens at first, or a disagreeable sensation of heat be felt in the mouth and throat, this is of little consequence, and such secondary inconveniences should be neglected. During the course of the malady the dose of alcohol, cinchona, or syrup may perhaps be modified, but this is the only change which need occur, and the draught containing wine and spirit should invariably be employed until the end of the disease.

I have never met with any serious difficulty in continuing these remedies in their fundamental part, namely, wine, cinchona, and brandy, but some circumstances should be mentioned which oblige an alteration to be made in the composition of the mixture. Certain patients, after taking it for a few days, complain that they cannot tolerate the treatment so well; first eructation takes place, then a dislike is felt for the remedy, and if it is continued, vomiting may take place from time to time. By omitting the tincture of cannella, this inconvenience is usually caused to disappear. In other cases intolerance of a different kind supervenes. While continuing to take, voluntarily, the remedy, the patient complains of the thick consistence which the extract of cinchona gives to the liquid, and before long the distaste results in vomiting. There is then no reason to insist upon this remedy, and the extract may be omitted; but since it is essential to administer cinchona bark with the principles contained in it, wine containing this remedy may be given in the place of that used as a vehicle, Bordeaux, Malaga, Madeira, or sherry being chosen according to the taste and constitutional condition of the patient.

When the disease is already of long duration, an erythematous form of stomatitis is not uncommonly observed, characterized by removal of the epithelium, denudation of the papillæ of the tongue, bright red colouration of the lining membrane of the mouth, and a most painful sensation of burning heat. This condition is not due to the stimulating treatment, and may be observed in those to whom this treatment had not as yet been administered. It may form, however, a real impediment to the administration of alcoholic remedies, which would at once increase the sensation of burning in the affected parts. The condition of the mouth is, therefore, to be carefully examined, and when the first signs of stomatitis appear, lotions should be applied frequently on each day of decoction of barley or marsh-mallow, to which syrup of poppies (*sirop diacode*) has been added, in the proportion of one and a half to two ounces (40 à 60 *grammes*), in nine ounces (250 *grammes*) of liquid, and a small quantity (half a drachm to a drachm) of chlorate of potash. By means of this simple precaution I have always been able to continue the treatment without difficulty or discomfort to the patient.

It has been already said that the degree of pyrexia is one of the elements which should be taken into consideration with respect to the amount of alcohol given, the diminution or still more the cessation of fever imposing reduction in the amount of the remedy. It would, however, be a serious mistake to omit the remedy altogether in a sudden way. By doing this the symptoms indicating weakness will undoubtedly reappear, the organism not being deprived with impunity of this power ful stimulant. It is by degrees, by imperceptible modifications from day to day, that the alcohol should be suppressed, and in those cases which pursue so fortunate a course that actual convalescence seems to occur, this remedy should be continued during its whole existence in gradually decreasing amounts. In the same way, when the disease becomes chronic, and the treatment of ordinary phthisis can be substituted for the specific remedies of the acute period, alcohol should be administered in larger quantity than when the illness is chronic from the moment of its onset.

While an endeavour is thus made to fulfil the fundamental indications drawn from the condition of the patient as regards strength, and the consuming nature of the disease, the pyrexia, which, as has already been fully explained, is itself a process of wasting, must be unceasingly opposed. Otherwise, if continuing for some time without notable remission, it will cause actual destruction of the most indispensable organs and tissues, notably of the heart. In the conditions which are being considered, it is impossible to remove permanently all liability to fever. Apyrexia will occur when local increase of the disease ceases to take place, as rarely happens, or when it passes into a chronic condition, as is more frequently the case. By means, however, of well-directed antipyretic treatment, which is perseveringly continued, real and useful control may be obtained over the febrile symptoms. In reality the monotonous pyrexia may be thus controlled at will, and as often as is thought necessary, and the patient can repeatedly obtain the benefit of decided remissions lasting for three or four days, which may be regarded as times at which the organism ceases to be consumed by the febrile heat, and the more so since its degree and continuity cause it to be far more dangerous than in the chronic form of phthisis.

In my opinion salicylic acid should be used, being replaced by hydric bromate of quinine if ill tolerated, and subcutaneous injections of the same salt should be made when the existence of dyspepsia contra-indicates the continuance of the acid. The plan which I usually adopt is to treat the patient during three days; on the first day the maximum dose is given, on the second a dose containing seven grains less, the same reduction being made on the third day. As regards salicylic acid, thirty grains are administered on the first day, twenty on the second, and fifteen on the third. Of quinine thirty grains are given on the first day, a similar reduction being made on the two following days; and lastly, when injections are used, fifteen grains should be taken on the first day, seven grains on the second, and third, etc., according to the rules previously laid down with respect to the hours and means of administration. Reduction of the temperature to a greater or less extent always occurs, lasting from two to five days; not that the lowest temperature exists during the whole of that time, but this or even a longer period of time passes before the temperature returns to the same point as before. With respect to subsequent treatment, the plan which I adopt is the following. From the time of first treating the patient stimulants are given, but an interval of fully two days is allowed to pass before antipyretic remedies are administered, so that an idea may be obtained as to the intensity of the fever when left without special treatment. On the third day antipyretic agents are employed, and continued for three days, as I have already explained, and the same plan of treatment is recommenced whenever the temperature reaches, either in the morning or evening, the same or nearly the same height as when the first observation was made. In this manner the patient has in the mean three days out of six during which no remedies are administered when antipyretic remedies are in use, and the fever is incessantly controlled and kept at the minimum degree which is compatible with the nature of the affection. It is after comparing for a long period the different modes of treatment that I feel able to recommend this plan, as that which unites to the fullest extent a powerful effect with facility of toleration.

As before observed, a special indication of digitalis may exist for a time during the course of the disease, as will be explained, and experience confirms the truth of ideas already mentioned in my Clinical Lectures, and which need only be repeated. In the different conditions which produce phthisis, which though acute are always of long duration, the contractile energy of the heart is necessarily compromised by the existence of fever as in every disease of long duration, while to this common cause of cardiac debility a particular one is added, to which sufficient attention has not been paid. The alterations in the lung are frequently of sufficient size and depth of position to embarrass the flow of blood in the pulmonary artery. When the destruction of tissue has commenced, this effect is increased by changes which occur in some branches of that vessel, and these conditions produce overfilling of the right ventricle, which crushes, as it were, the already failing force of the heart. In this condition, which the weak impulse and heart-sounds, the feeble, small, compressible, and often intermittent pulse, the cyanosis of the face and stasis of blood in the cervical veins reveal to exist, not a moment should be lost. The action of stimulants which affect the whole organism is not enough, and the contraction of the heart must itself be directly excited. This indication is evident and urgent, and the means of fulfilling it are supplied by digitalis.

When the heart first shows signs of failing, the employment of quinine and salicylic acid should give place to that of digitalis. Small doses should be given on account of the general state of the organism, of which the enfeebled condition might cause a poisonous to be substituted for a therapeutic effect. For the same reason the remedy should be omitted whenever the contractile power of the heart is sufficiently restored to remove all immediate danger, and be recommenced should the heart again fail. In my opinion the infusion made from leaves recently crushed should be used, the amount of leaves employed daily varying from five to twelve grains. The infusion may be made in four ounces of water, and the liquid sweetened with sugar so as to become a remedy which can be employed alternately with that containing alcohol, which should be rigidly maintained. I prefer the plan of having an infusion of digitalis made with a very small quantity of water, six ounces, for instance, and adding this liquid, when thoroughly filtered, to the alcoholic drink. Two advantages are gained by this method, that of avoiding any complication in the treatment adopted, and of avoiding as far as possible the nausea and vomiting which digitalis is the more likely to provoke in proportion as the weakness is more pronounced.

This completes the number of internal remedies which I employ, and some of the most brilliant successes of my practice are due to them. At the same time, however, and with no less perseverance, I endeavour to act upon the local condition, and to prevent the congestive movements which increase the extent of the lesions by restraining the formation of fresh exudations, and promoting the absorption of liquid products. This triple effect I endeavour to obtain by means of large flying blisters, without leaving, as it were, a moment unemployed in the use of counter-irritants. When once the part has almost or altogether healed, I cause another blister to be placed in the adjoining region, so that the part may be in constant activity. Should there be decided evidence of vesical irritation, I cease to continue the application of blisters, but encourage the suppuration of a blistered part placed upon the anterior surface of the thorax. This plan, though apparently severe, specially when the uncertainty of the result is considered, may still be considered beneficial when conforming to the preceding rules.

Such treatment is continued with but slight modification during the whole of the acute period. If the disease fortunately passes into a chronic stage, the condition is before long identical with that which exists when similar lesions occur in the ordinary form of phthisis; it is, therefore, unnecessary to say more about the pneumonic form.

Acute miliary tuberculosis will be now considered.

This, without doubt, is the most formidable of all forms of phthisis, death occurring as a rule within a few weeks. At the same time, however incredible the statement may appear, exceptions undoubtedly occur—rare, it is true, but which should be carefully registered, firstly on account of their existence, secondly by way of encouragement. These exceptions do not occur either in the suffocating or catarrhal form of the complaint, belonging to the disease which is of longer duration, even when fatal, and which is known as the typhoid form of phthisis. I mentioned, however, in my Clinical Lectures the four cases of Lebert, which prove that recovery from this affection is not anatomically impossible. At the autopsy of four persons who had succumbed to other diseases he found traces of disseminated miliary tuberculosis which had ended in recovery, the lesions produced by the granular production presenting the characters of a process which had completely ceased. The same observer quotes two other cases, in which he observed clinically the cure of the disease, the special course of the illness leaving no doubt as to the diagnosis of the malady. On the other hand Sick, and more recently Anderson, have called attention to similar facts, so that though the number of cures recorded may, as I admit, be too small to authorize the existence of much hope, yet the opposite conclusion, which until the present time has been considered as certain, should now be suspended.

I feel the more authorized to make this consoling reserve from having myself observed in 1879 an undoubted example, as I believe, of complete and definite recovery from diffused miliary tuberculosis. The subject is so important that I feel no hesitation in mentioning the most striking and certain particulars of this case.

A girl, Marie B——, aged sixteen years, was admitted under my care March 6, 1879, on the sixth day of a serious febrile disease. The appearance of the patient, the evening temperature at the time of her admission, caused the disease to be regarded as a most serious form of typhoid fever, a diagnosis which was maintained for forty-eight hours. After this time the absence of abdominal symptoms of epistaxis, of rose-coloured spots, of any remission in the temperature from the sixth to the eighth day, and the existence of oppression which was quite disproportionate to the few sonorous *râles* heard in different parts of the chest, caused legitimate doubt to be felt as to the truth of the above diagnosis. On the tenth day of the disease this diagnosis was unhesitatingly abandoned, and acute disseminated granulosis was believed to exist, since over the pleura anteriorly and at the lower part pleural friction was heard on both sides, while some effusion was found to exist posteriorly on the left side. At the same time that diarrhœa became established, and the abdomen distended with gas, intense pain was felt throughout the abdomen, the maximum of which was always clearly referable to the perihepatic and perisplenic regions. In the presence of these pleural and peritoneal symptoms doubt was impossible, and the newly formed conclusion was confirmed two days later, when, without extension of the pleural friction, symptoms of bronchial catarrh were found to be marked over the upper lobe on each side. The fever persisted, specially notable on account of its continued irregularity, and there was emaciation, which reduced the patient with such surprising rapidity that before the end of the third week a condition of actual marasmus might be said to exist. The disease continued to progress with but slight variations during three weeks, the evening temperature remaining at from 102° to 104° F. (*entre* 39° *et* 40° *C.*) except when antipyretic remedies were used. Notwithstanding the pronounced character and duration of the pyrexia, the heart-sounds remained of the same intensity; the only indication of brain affection was slight delirium in the evening and at night, due to the existence of fever, while no symptom was at any time observed indicating the formation of tubercles in the brain. The stethoscopic signs furnished by the lungs never showed the existence of more than disseminated bronchial catarrh. Crepitant *râles* of variable size in abundance, and without notable diminution of resonance, were alone observed, no evidence existing at any time of consolidation or hepatization. After the 5th of April the character of the fever changed. While there was no pyrexia in the morning, the temperature of the evening was often as high as 102° or more, and on the 12th of May, the seventy-fourth day of the disease, the absence of evening pyrexia was the signal of commencing convalescence. The time of recovery was proportional to the severity of the affection and the great intensity of the organic loss; the improvement continued without notable change, so that on the 30th of July the patient was able to leave the hospital entirely cured, after remaining there for nearly five months. The only remains of the affection were adhesions of the pleura to the diaphragm, and a scarcely perceptible diminution of resonance in the semilunar space.

At the end of June, six weeks after complete cessation of the fever, her weight was 4 st. 3 lbs. (29 *kilogrammes*) ; when she left the hospital, at the end of July, it was 4 st. 11 lbs. (33 *kilogrammes*); when seen again in August the weight had increased by 6 lbs. (3 *kilogrammes*), and in September was as much as 5 st. 5 lbs. (38½ *kilogrammes*).

Five months and a half later, on Jan. 15, 1880, I saw her again, and could once more realize the completeness of the recovery. The strength of her constitution had improved considerably, the weight being now 5 st. 10 lbs. (41 *kilogrammes*), this being an increase of 5 lbs. (*deux kilogrammes et demi*) since the month of September.

This is a remarkable case, which, though I have observed no other, seems sufficient, even if unique, to establish at least the possibility of absolute recovery in acute miliary tuberculosis.

To include every favourable result which this form of tuberculosis may present, another possibility should be taken into consideration, which, though by no means so favourable as the preceding, is still undoubtedly fortunate in this disease. The following takes place. After some weeks the disease becomes less acute, the pyrexia ceases, and the complaint passes slowly into a chronic illness; that is to say, the acute miliary disease changes into the ordinary form of granular phthisis. I have observed three examples of this alteration, which, as regards the rapidly fatal issue of the disease, is truly a relative cure. The three cases were each of the prolonged form of the disease termed typhoid, by which change alone, as I believe, is such a result possible. I need scarcely add that in these patients there was no evidence of brain disease, as is so often the case in diffused miliary tuberculosis.

To sum up, complete recovery, or transformation into a chronic disease, should be regarded as possibilities which diminish to a slight extent the sentence of absolute incurability which has been pronounced against the acute form of miliary tuberculosis.

In order to increase this chance of a favourable result, what course should be adopted? After the great length at which pneumonic tuberculosis was considered, a few words will enable the principles and practice of the treatment which I adopt in such circumstances to be fully understood. Without being engrossed by the tuberculous nature of the lesion, and the fact of its being usually incurable, I treat the acute miliary disease during its whole course as an infectious complaint, or more precisely, as a typhoid fever with hyperpyrexia. In my opinion, the three fundamental indications which I perseveringly attempt to fulfil are to remove the pyrexia, to lessen the dyspnæa and improve the local lesions, and to sustain the strength. I fulfil the first indication by means, not of sulphate, but of hydric bromate of quinine, a remedy which joins to the antipyretic action of quinine salts an antiseptic or anti-fermentative effect, as it may be termed, this being of the greatest importance in a morbid condition, the danger of which is usually increased sooner or later by the fact that absorption is taking place at the seat of the local lesions. That repetition may be avoided, I will not again dwell upon the details and modes of applying the appropriate remedies, which have already been fully considered. To fulfil the fundamental indication, I apply a lotion composed of aromatic vinegar, in a pure state or combined with cold water, from four to eight times daily, as in typhoid fever. Baths should not be recommended owing to the intensity of the dyspnæa, the importance of the lesions in the chest, and the frequent existence of abdominal complications.

The second indication is to lessen the dyspnæa and improve the local lesions. This I fulfil by means of repeated application of large blisters over different parts of the chest, specially over the recognized foci of disease. When the dyspnæa is serious, and the bronchial catarrh extends over a large part of the lung, I also employ the process of dry-cupping, from forty to sixty cupping-glasses being placed in the morning and evening over the lower extremities and trunk. If, in the last place, peritoneal complications occur of some intensity, as indicated by the pain, flatulence, and partial or total immobility of the diaphragm, ice is applied permanently to the anterior surface of the abdominal wall, a process which is certainly inconvenient, but which experience has shown me to be of invaluable benefit.

To maintain the strength is a third indication the importance of which need not be dwelt upon. Wine, alcohol, in quantity proportional to the age, the constitution, and habits of the patient, as well as the weakness and degree of fever, cinchona bark, nourishing food, consisting of broths, jellies, milk or peptone, afford the means of fulfilling this obligation which is so truly vital.

Such is the treatment which I employ in acute miliary tuberculosis, and to which I owe the results already mentioned, and I am disposed to believe that a favourable termination would be more usual if the physician was less inclined to be discouraged and inactive when in the presence of this formidable disease.

CHAPTER X.

TREATMENT BY MINERAL WATERS.

Utility of mineral waters in the treatment of pulmonary phthisis—Limit to their use and mode of action—General counter-indication to their employment.

Indications—Special difficulties of the subject—Complexity of the elements of decision — Successive stages of judgment — Individual character of the therapeutical decision.

Prophylactic period—Sources of the indications—Means of fulfilling them.

Ordinary phthisis—General counter-indications—Indications—Reaction of the patient against the disease.

Primary forms of phthisis—Sources of the indications.

New method of classifying mineral waters—Division into three groups—Applications.

Secondary phthisis—Scrofula—Arthritism—Herpetism—Pneumonic phthisis.

THAT no mineral water has any action upon tuberculous products is a proposition which undoubtedly contains nothing new, but expresses a fact, unfortunately, but too certain and well known. Such being the case, is the conclusion which is made by some physicians justifiable, that mineral waters are useless in the treatment of pulmonary phthisis, and may be omitted without detriment to the patient? Not altogether, in my opinion; such a reason would lead to the omission of all treatment, since none of the means already considered at such length, and of which the employment has been so strongly recommended, has any effect upon the tuberculous products. At the same time we have recourse thereto, as it is our duty to do, since, though not removing the tubercle, they benefit the patient and the condition of the lungs.

The same may be said of mineral waters; without effect upon the tuberculous lesions, they may be for other reasons of threefold utility.

Firstly, they affect the general condition of the patient and the activity of nutrition, that is, the constitutional malnutrition, which is the fundamental pathogenic condition of the disease. This effect they have, not only owing to their composition, but also on account of the climatic and hygienic conditions associated therewith by which this effect is produced.

Secondly, they act with as much effect upon the morbid states of the constitution, which but too often accompany the development of phthisis.

Lastly, they may arrest the perituberculous lesions, that is to say, the pneumonic, broncho-pneumonic, or catarrhal affections, which in most cases increase the extent of pulmonary change, aggravate the condition of the patient, and are the means, not to say the condition, of extension of the tuberculosis. By reducing these contingent lesions to a minimum degree, by preventing their return for a greater length of time, a signal service is rendered to the patient, and a result of great importance is obtained. Though the mineral waters are not the sole remedy which can have this effect, it is certain that they do so if judiciously employed, and that reason alone would be sufficient to justify their use, and even to impose it in certain definite conditions.

Mineral waters may be most useful in another way, and this advantage, which may at first be considered accessory and commonplace, is in my opinion real and important to the patient, when necessity or blind opposition to advice induces him to remain in town. The effect of this treatment is to remove the person for some weeks at least from the confinement and bad hygienic surroundings which aggravate and precipitate the disease, obliging him to live in the open air, while to this healthful physical influence a no less fortunate moral effect is added, which prevents discouragement, and in my opinion none of these considerations should on any account be disregarded by the physician.

Such propositions, which experience has shown me to be true, indicate the exact degree to which mineral waters may be considered beneficial in the treatment of phthisical patients. It would be unfair to conceal the opposite truth, namely, that these waters may be injurious, a fact which at any rate shows that as medicinal agents they are not without effect. Thus harm may be done when they are not appropriate to the condition of the patient and the form of the disease, when they are administered to an unreasonable extent, or at such an advanced period of the complaint, that any change of residence is dangerous, as a probable source of fresh complications, and when the fatigue necessary to recovery is not compensated by the results which may be expected. These unfortunate consequences, however, are always the sequel of a mistaken or ill-timed use of the remedy, and do not in any way disprove its inherent advantage when properly administered.

Thus mineral waters should, in my opinion, be looked upon as a means of treatment which may be beneficial in pulmonary phthisis, but which is not necessary in every form of the disease. This entirely depends upon the indications which exist.

These indications will now be considered, those alone being named which direct my own personal practice, the course always pursued in these observations. Without wishing in any way to criticize or compare, I would only mention my own plan of treatment, and the reasons why it is adopted in preference to others.

Before more is said, two conditions should be mentioned which though foreign to the pathology of the complaint set a limit, as I think, to their use. These I invariably consider in the practical treatment of the disease, namely, the kind of life led, and the course of treatment followed by the patient. Whether during the prophylactic period or when the disease exists, should the patient live in the mountains, or in country air, where the climatic conditions are in every respect appropriate to his condition, if he is pursuing the hygienic, hydro- and aërotherapeutic means of treatment already mentioned, at the same time that the internal and external medical agents are being regularly employed, no modification should be made in a mode of life which, in my opinion, is the most advantageous possible. The employment of mineral waters at a distance may then be abandoned without hesitation or regret, the certain effect of their use being to cause abandonment of most beneficial habits. If in this disease the constitutional condition of the patient seems to indicate that one of the mineral waters would be beneficial, it can then be taken without change of residence. The effect of the remedy is diminished, but at any rate no harm will have been done by substituting a certain good for a doubtful better.

It should be observed that this special counter-indication rarely occurs. It is but very few patients that can remain in these favourable conditions, specially as regards a country life. Not having the means to do so is the chief obstacle, while the varying disposition which exists in this class of patients is a second difficulty, no less often encountered, and as much to be taken into consideration. In consequence the above-mentioned counter-indication is but exceptional, and in the reverse conditions of life, which in fact are usually met with in practice, the indication of mineral waters should be fulfilled if it exists. Hence the proposition which I have just made, that mineral waters are a means of treatment which may be beneficial, but which should not be employed unreservedly in the case of every patient. The above considerations explain and complete my views upon this subject, which I would still illustrate by some examples.

The patient who will first be considered is in that condition when living at a high altitude is the best treatment, and he consents to remain for a year or more, if necessary, at Davos* and in the Engadine,† employing the hygienic and medical remedies which are specially appropriate to his condition. Leaving an atmosphere, which is the fundamental element of treatment in his case, for the sake of mineral waters, might cause him to lose in a few weeks the improvement gained after many months. Another patient, who specially requires at this moment a mild and equable climate, has consented to remain for one or two years in one of the few places which are appropriate during the whole year—as, for instance, Madeira.‡ In this case there would be no advantage

---

* Davos is a district in the canton of the Grisons, Switzerland, Davos am Platz, its capital, being much frequented by invalids. See *Lancet*, Oct. 17, 1877; "Davos Platz, a New Alpine Resort for Sick and Sound in Summer and Winter" (1878), E. Stanford; *Fortnightly Review*, July, 1878, Article by J. A. Symonds; *Fortnightly Review*, Nov., 1879, Article by J. Burney Yeo, M.D.; "Dictionary of Watering-Places" (L. Upcott Gill) "Winters Abroad," R. H. Otter; "On the Climate of Davos am Platz," Alfred Pope; "Health Resorts," J. Burney Yeo, M.D.; "Principal Southern and Swiss Health Resorts," Wm. Marcet, M.D.; "Handbook for Switzerland" (J. Murray); "Handbook for Travellers in Switzerland," K. Baedeker; "Hachette's Diamond Guide to Switzerland," A. and P. Joanne; "Influence of Climate on Pulmonary Consumption," C. J. Williams; "The J.E.M. Guide to Davos Platz," J. E. Muddock; "The Alps and how to see them," J. E. Muddock; etc.

† The Engadine, or valley of the River Inn, extends from Maloja to near Zernetz (Upper Engadine), and from thence to the frontier of Tyrol (Lower Engadine). See "Handbook for Travellers in Switzerland" (J. Murray); "Guide to Travellers in Switzerland," K. Baedeker; "Hachette's Diamond Guide to Switzerland," A. and P. Joanne; "The Alps and how to see them," J. E. Muddock; etc.

‡ Madeira is the island in the North Atlantic Ocean off the west coast of Africa. See "Curative Effects of Baths and Waters," Julius Braun; "Memoirs of Life and Work of C. J. B. Williams;" "Health Resorts," J. Burney Yeo, M.D.; "Principal Southern and Swiss Health Resorts," Wm. Marcet, M.D.; "Handbook of Madeira," J. M. Rendell; "The Climate and Resources of Madeira," M. C. Grabham; "Madeira, its Climate and Scenery," R. White; "The Climate of the South of France," etc., C. J. Williams; "Influence of Climate on Pulmonary Consumption," C. J. Williams; "A Visit to Madeira in the Winter of 1880-1881," D. Embleton, M.D.; "Madère étudiée comme Station d'Hiver," J. Goldschmidt; etc.

in the use of mineral waters, whereas some relapse might occur, which experience has shown me to be often the case.

After what has been said, the meaning and extent of this counter-indication will be quite understood, being simply, in fact, an application of the old maxim, *primum non nocere*. If the conditions of the patient as regards hygiene, climate, and treatment are favourable; if the residence is appropriate, and he consents to remain therein during the time required that the treatment may be effectual, any removal therefrom would be injurious, by causing the healthful habits which had been adopted to cease, and the use of mineral waters at a distance is thus counter-indicated.

This being perfectly understood, the indications of mineral waters will now be considered. It may be unhesitatingly said that this question is one of the most difficult, for reasons which will be stated, and the knowledge of this fact should induce caution, prudence, and circumspection. Nothing is more difficult than to form a conclusion in therapeutics, but in addition to the difficulty which exists in all such decisions, and which specially regards the patient, another, resulting from the complexity of the data, should be taken into consideration. The composition of the water, that is as regards its nature and effect, is but one element in the decision formed. It is certainly the only one usually considered in such a case, but to take such a narrow view of the question is a mistake which does harm to the patient, and brings discredit upon the use of mineral waters. In order to avoid this twofold error, the climatic conditions of the place selected should be considered with the same forethought. The mineral water itself is but one means of action, while the climate is another, which has a good or bad effect according as it is favourable or unfavourable to the condition of the patient, and it is a real mistake to regard an agent of such value in the treatment by mineral waters as of secondary importance. In some cases the importance of the climate is such that it surpasses, or should surpass, that of the water's chemical composition. It is my habit never to separate these elements, as will be shown in the following pages, and in mentioning the two fundamental facts all is not yet said, since the sanitary condition of the place must be taken as seriously into consideration, with regard to affections which are either endemic, or happen to prevail when this form of treatment is to be employed. The therapeutic resources of the place must also be considered with respect to baths, shower-baths, inhalation of spray, and pneumatic machines. Finally, attention should be paid to the possible dwelling of the patient, as regards its position and the power of obtaining food. If any alteration is to be made in these, care should be taken that it is an improvement, and not the reverse. Should the residence be inappropriate, and the patient be suddenly deprived of comforts to which he is accustomed; if, on account of the small size of the place, baths must be taken at inconvenient hours, which necessitate an abridgment of sleep; if the process of inhalation must be carried on by the patient in the company of others—a practice which I strongly condemn, both from the patient's repugnance to it and the danger which it involves—this plan of treatment, however strongly indicated in other ways, will undoubtedly be injurious to the patient.

Thus it is not sufficient to classify the mineral waters in divisions founded upon analogy of composition;* if this alone were required, the task would be comparatively easy, since there is no excessive difficulty in thus establishing chemical or pathological groups. Much more, however, is necessary, since, owing to the numerous above-mentioned elements upon

* The classification usually made depends upon the chemical composition of the mineral water. See "The Elements of Materia Medica and Therapeutics," J. Pereira, M.D.; "The Essentials of Materia Medica and Therapeutics," A. B. Garrod, M.D.; "Nouveau Dictionnaire de Médecine et de Chirurgie Pratiques, Paris;" "The Baths and Wells of Europe," J. Macpherson, M.D.; etc.

which the decision is founded, a complete knowledge should exist of the different units which form the total. Even when this is the case, though some indication will have been acquired, the solution of the problem is not yet obtained. The most beneficial place in the group selected must be chosen with equal care, and this second step is of no less importance than the first. Thus in many cases it may be as harmful to choose an individual place which is ill adapted to the case, as one which belongs to the wrong chemical group. If, for instance, sulphuretted waters which contain soda are considered beneficial to the patient, and, without other circumstances being taken into account, he is sent to Amelie-les-Bains * when his organism requires the climatic conditions of Cauterets,† more harm, or at any rate less good, will be done than by mistaking the group of mineral waters and sending him to La Bourboule,‡ which is at any rate of the same altitude.

The diversity of and logical connection between the successive points to be settled in making the choice of a mineral water will now be discussed. The first question to be decided is the chemical group to which the mineral water should belong. This is ascertained on the one hand from the

* Amelie-les-Bains is a French watering-place, department Pyrénées Orientales. See "Baths and Wells of Europe," J. Macpherson, M.D.; "Health Resorts," J. Burney Yeo, M.D.; "Medical Guide to the Mineral Waters of France," A. Vintras, M.D.; "Principal Baths of France, Switzerland, and Savoy," Edwin Lee, M.D.; "Curative Effects of Baths and Waters," Julius Braun; "Handbook for Travellers in France" (J. Murray); "Dictionnaire Encyclopédique des Sciences Médicales," Paris; "Guide to the South of France," C. B. Black; etc.

† See note, p. 24.

‡ La Bourboule is a French watering-place in the arrondissement Oloron, in the Basses Pyrénées. See "The Baths and Wells of Europe," J. Macpherson, M.D.; "Medical Guide to the Mineral Waters of France," A. Vintras, M.D.; "Health Resorts," J. Burney Yeo, M.D.; "Curative Effects of Baths and Waters," Julius Braun; "Handbook for Travellers in France" (J. Murray); "Dictionnaire Encyclopédique des Sciences Médicales," Paris; "Auvergne, its Thermo-mineral Springs," etc., Robert Cross, M.D.; "Guide to the South of France," C. B. Black; etc.

patient, that is to say, by learning his family history, his antecedents, and the course of the existing complaint; on the other, from information acquired by experience as to the adaptation of certain mineral waters to special local or constitutional pathological conditions, and to the reaction of the individual patient affected by the disease. When these data are acquired, the place belonging to the group selected must still be chosen, and this is done by acquaintance with the patient, his constitution, his tendency to excitement, habits, residence before and after such treatment, the result of previous change of residence, the symptomatic particulars of the disease, the condition of the heart and large vessels on the one hand, and on the other by more or less complete acquaintance with the different places of the group, specially as to their climatic conditions, and the more or less stimulating effect of the water. It is then possible to form a conclusion, and, in my opinion, no decision connected with the therapeutics of disease requires so complex an intellectual process, so delicate a tact, and such extensive knowledge. The task is much facilitated by a direct knowledge of the place, of which fact I have been convinced ever since the beginning of my career.

The result of what has been stated is that the choice of mineral waters has a purely individual character, and that no general rule can be made. With regard to the first part of the problem, the most decided forms of the disease, and the fundamental types of organic reaction, may certainly be taken into consideration, and corresponding groups be formed accordingly. The relations established between groups formed according to the chemical composition of the water and pathology of the complaint may be thus expressed in general terms. As regards, however, the second part of the problem, any attempt to do this would be totally useless. The indication is individual in a double sense, since not only must the individuality of the patient be considered as in all cases but also that of the different stations of the group. With such complexity of data no general rule can possibly be made. It is true that some of the facts are constant, or nearly so, namely, those which concern the individual watering-place; others, however, which are of no less importance, undoubtedly and continually vary. Differing in each patient, the treatment which is beneficial in one case is not so in another, and the whole problem must be again and again solved. Completely new facts present themselves repeatedly for medical appreciation, the solution being indicated by no precise law, but retaining in all cases a unique and exclusively personal character.

It will thus be understood that, in the following remarks, the indication of groups rather than of individual watering-places will be considered. As far as possible reasons will be given why one or other of the principal stations in the group should be preferred, this, however, being necessarily done in the briefest manner; and with regard to individual patients, one may at any time be led to transgress rules that have been made, and the elements of therapeutic management, since the selection must depend upon the person concerned, and is not susceptible of didactic expression.

The details of this subject will now be considered.

The prophylactic period, the importance of which I have endeavoured to show in the preceding chapters, is without doubt that in which the mineral waters can render the most signal and durable service, the hygienic and climatic treatment in its completeness not being then possible. Three groups of cases must be considered. The first consists of those in whom, owing to hereditary tendency, personal cause, or other reason, the ultimate development of tuberculosis is to be feared, and who present no distinctive indication other than the debility or constitutional malnutrition which is the common basis of all forms of phthisis, the organism being free from any appreciable pathological lesion. In the second group, the constitutional debility is accompanied by obvious and persistent deficiency of blood-globules. In the third and last group, the suspicious constitutional condition is associated with existing or previous manifestations of scrofula.

In the first group of cases, which, however, is the least numerous, mineral waters are useless, so long as all the details of the treatment which have been already recommended at that time are scrupulously observed, as regards residence, hygiene, physical exercise, and medicine. These measures will produce, in my opinion, all the improvement which it is possible to effect. When, however, this is not possible, as, for instance, with the inhabitants of towns, mineral waters are beneficial and even indicated, notwithstanding the absence of any definite pathological lesion, and the vague meaning of the expression "constitutional debility." The chief advantage of the waters is then of an indirect character, though no less valuable on that account. Their effect is to produce for a time a result which could not be obtained by adopting other means of treatment, while a pretext then exists for making a total change in the regulations of life; the defective hygiene of a town is replaced by the health-giving habits of the country, and although but temporary, these changes undoubtedly tend to promote the organic repair which is so specially required.

Mineral waters themselves tend also to produce such a result, since in these particular cases the selection of the water must be made from three groups, which, independently of their special action, have in each case a strengthening and tonic effect. Thus the choice will be between ferruginous waters, those with chloride of sodium, and those which contain sulphurous acid; and in the absence of special indications furnished by anæmia or scrofula, the choice will depend upon the antecedents and constitution of the patient, the family history being also taken into consideration. When the group is selected, the choice of the individual place seems at first to be totally free, owing to the absence of any indication on the part of the patient, and the use of mineral waters being but a detail in the treatment. This, however, in my opinion, is by no means the case. The treatment is now prophylactic, and the constitutional strength must be restored, so that on this account the climatic conditions are, as I think, quite as important as the composition of the water. Thus, that no means of action may be neglected, a place should be chosen of which the climate is most apt to promote activity of nutrition and increase the strength of the organism. Altitude is the fundamental condition of such climatic action, and in consequence the most elevated stations in each group should, in my opinion, be chosen so long as the heart and nervous excitability present no counter-indication. St. Moritz * in Switzerland, of the ferruginous, Gurnigel † in Switzerland, Cauterets ‡ and Allevard § of the sulphurous

---

* St. Moritz is a town in the Upper Engadine, Switzerland. See "Dictionary of Watering-Places," part ii. (L. Upcott Gill); "Notes of a Season at St. Moritz in the Upper Engadine" (1870), J. Burney Yeo, M.D.; "Health Resorts," J. Burney Yeo, M.D.; "Principal Southern and Swiss Health Resorts," Wm. Marcet, M.D.; "Health Resorts for Tropical Invalids," Moore ; "Curative Effects of Baths and Waters," Julius Braun; "Handbook for Switzerland" (J. Murray); "Guide to Switzerland," K. Baedeker; "The Alps and how to see them," J. E. Muddock; "European Guide-book," etc., Appleton; "Hachette's Diamond Guide to Switzerland ;" etc.

† Gurnigel is a Swiss watering-place in the canton of Berne. See "Health Resorts," J. Burney Yeo, M.D.; "Curative Effects of Baths and Waters," Julius Braun; "Principal Southern and Swiss Health Resorts," Wm. Marcet, M.D.; "Switzerland and the Adjacent Portions of Italy," K. Baedeker; "Handbook for Travellers in Switzerland" (J. Murray); etc.

‡ See note, p. 24.

§ Allevard is a small French town, in the department of the Isère. See "Medical Guide to the Mineral Waters of France," etc., A. Vintras, M.D.; "The Baths and Wells of Europe," J. Macpherson, M.D.; "Handbook for Travellers in France," part i. (J. Murray); "Dictionary of Watering-Places" (L. Upcott Gill); etc.

waters, La Bourboule* and Salins† of the group which contains chloride of sodium, and Uriage,‡ the waters of which contain both sulphurous acid and sodic chloride, are the stations which seem preferable from this point of view.

The second group of facts which should be distinguished in the prophylactic period presents a more precise and limited indication. Acquired or innate constitutional debility is associated with deficiency of blood-corpuscles. Should this condition be unaffected by the treatment appropriate to the disease, or if the treatment cannot be practised in its entirety, as already indicated, recourse should be had to mineral waters containing iron, of which the special action may be powerfully aided by the climatic conditions of certain places. From this point of view St. Moritz§ and St. Catherine ‖ are typical stations, while Forges-les-Eaux,¶ Spa,** Schwalbach,††

* See note, p. 246.

† Salins is a French town, in the department of Jura. See "Medical Guide to the Mineral Waters of France," etc., A. Vintras, M.D.; "Dictionnaire Encyclopédique des Sciences Médicales," Paris; "Handbook for Travellers in Switzerland," K. Baedeker; "Hachette's Diamond Guide to Switzerland;" etc.

‡ Uriage is a French watering-place, in the department of the Isère. See "The Baths and Wells of Europe," J. Macpherson, M.D.; "The Mineral Waters of France," etc., A. Vintras, M.D.; "Handbook for Travellers in France" (J. Murray); "Curative Effects of Baths and Waters," Julius Braun; "Guide to the South of France," C. B. Black; etc.

§ See note, p. 250.

‖ St. Catherine is a Swiss watering-place, situated in the canton of the Valteline, about seven miles from Bormio. See "Health Resorts," J. Burney Yeo, M.D.; "Handbook for Travellers in North Italy" (J. Murray); etc.

¶ Forges-les-Eaux is a French town in the department of the Seine Inferieure. See "Medical Guide to the Mineral Waters of France," A. Vintras, M.D.; "Handbook for Travellers in France," part i. (J. Murray); "Dictionnaire Encyclopédique des Sciences Médicales," Paris; etc.

** Spa is a town in Belgium, in the valley of the Ardennes. See "Dictionary of Watering-Places," part ii. (L. Upcott Gill); "Curative Effects of Baths and Waters," Julius Braun; "Handbook for Travellers in Holland and Belgium" (J. Murray); "Handbook for Travellers in Belgium and Holland," K. Baedeker; "Dictionnaire Encyclopédique des Sciences Médicales," Paris; "A Three Weeks' Scamper through the Spas of Germany and Belgium," Sir Erasmus Wilson; "European Guide-book for English-speaking Travellers," Appleton; etc.

†† Schwalbach is in North Germany, being situated about fifteen miles from

and Pyrmont * should also be mentioned, although these places have no climatic action.

The exceptional altitude of St. Moritz, at which place the baths are 5810 (1770) and the town 6090 feet (1855 *mètres*) above the level of the sea, causes it to have a double therapeutic action. As elsewhere observed, while Schwalbach, Forges, Spa, and Pyrmont are only beneficial on account of the waters which they contain, St. Moritz is so on account both of its waters and climate, which may be truly qualified as having a tonic and stimulant character, the latter being also characterized by pronounced rarefaction of the atmosphere. The mean height of the barometer from the month of June to that of September is 24·3 inches (616 *mm.*), that is to say 5·7 inches (144 *mm.*) less than its mean height at the level of the sea,† whilst its extreme variations are comprised between 23·6 and 24·7 inches (599 *et* 627 *mm.*). These facts show that the climatic conditions of St. Moritz give to this place its therapeutic value, and it would be the greatest mistake to recommend this station from the sole fact that the patient's condition requires the employment of ferruginous waters. The least inconvenience of such practice would be that a long journey would be imposed on invalids, who perhaps, when the place was reached, could not manage to live there. What should be specially decided is the probable

Nassau. See "Dictionary of Watering-Places," part ii. (L. Upcott Gill); "Health Resorts," J. Burney Yeo, M.D.; "The Baths and Wells of Europe," J. Macpherson, M.D.; "Curative Effects of Baths and Wells," Julius Braun; "Handbook for Travellers in North Germany" (J. Murray); "The Rhine from Rotterdam to Constance," K. Baedeker; "Dictionnaire Encyclopédique des Sciences Médicales," Paris; "A Three Weeks' Scamper through the Spas of Germany and Belgium," Sir E. Wilson; etc.

* Pyrmont is a town of North-western Germany, in the Waldeck principality. See "Dictionary of Watering-Places," part ii. (L. Upcott Gill); "Curative Effects of Baths and Waters," Julius Braun ; "The Baths and Wells of Europe," J. Macpherson, M.D.; "Handbook for Travellers in Northern Germany," K. Baedeker; "Handbook for Travellers in North Germany and on the Rhine" (J. Murray); etc.

† More exactly the general mean at the level of the sea is 29·96 inches.

effect of altitude, and it is only when this question is settled that the employment of these ferruginous waters should be considered, though they are usually beneficial from containing a large quantity of iron and being easily digested.

Viewed in a general manner, the counter-indications to an altitude equal to or more than 5000 feet (1500 *mètres*) are numerous and of different kinds, but it would be leaving the special subject which is now being considered, namely, the treatment of a special group of patients, who are weak and anæmic, and in whom the development of tuberculosis is rightly dreaded, to speak of these facts. Within the limits above mentioned, the counter-indications are but few, though to be strictly respected. Lesions of the heart or large vessels, or pulmonary emphysema, absolutely prevent any benefit being derived from residence at high altitudes, and consequently at St. Moritz.

The functional disorders of the heart, and nervous excitability, so frequent in the anæmic, however pronounced in character, do not constitute any positive counter-indication, though they impose certain precautions with regard to residence at these places. If the patient is not willing, or cannot submit to these, the counter-indication is, in my opinion, positive, and some different arrangement should be made. It is requisite that the patient should become gradually accustomed to the altitude, a precaution, however, which is unnecessary or less important should he be inured to mountain air, while it is indispensable in the case of one who inhabits a valley, or is liable to the nervous and cardiac excitability already mentioned. Two stages should be made, the patient first becoming accustomed to the height of from 1500 to 2500 feet (*de cinq à six cents mètres*), and then to that of from 3500 to 4000 feet (*entre mille et douze cents mètres*). He should remain at each of these altitudes from three to five days, and it is only after this time that, having become gradually accustomed to the increasing rarefaction of the air, he can, without inconvenience, make his permanent residence in an extremely rarefied atmosphere. Should this precaution be omitted, the first days of such residence will undoubtedly be valueless to the patient on account of the rapid change imposed upon the organism by such a sudden transition, while usually the effect of this imprudence is still more unfavourable, since the patient, owing to sudden aggravation of all the troublesome symptoms from which he suffers, whether headache, insomnia, physical and moral excitability, palpitations, or difficulty in breathing, loses heart, and is so discouraged as to leave the station. Should he remain, the place having made an unfavourable impression upon him during the first days of his residence, he becomes prejudiced against it, shortens his stay to a minimum degree, looks upon his departure as a time of escape, and leaves at a time when, having become accustomed to the altitude, he might reap the benefits conferred by it. Thus such a plan fails in every way; and by neglecting a precaution which, in my opinion, is indispensable, the patient will have expended both time and trouble without obtaining the improvement which would undoubtedly have occurred had these precautions been taken. In other cases the cardiac troubles are aggravated during a long period of time, and the nervous exhaustion which results from the exciting influence of the air induces the most troublesome phenomena of anæmia, namely, those which result from such a condition in the brain and spinal cord.

There is one counter-indication, however, which remains, even though the above-mentioned measures be rigidly put into practice, namely, pronounced chlorosis, or anæmia. In these conditions, though mineral waters would be specially useful on account of the rapidity with which they induce the formation of blood-corpuscles, the climate, whatever precautions are taken, is too exciting, the organism not having the capacity of reacting, after so powerful a stimulant, either as regards the production of heat or improvement of nutrition. Such patients lose instead of gaining strength, become emaciated from having no power of repairing loss, and residence at the watering-place, if continued, becomes truly injurious to their health. This counter-indication has a purely individual character, being proportional to the powerlessness of the patient to withstand the known effects of the climate. Its value can only be measured by means of analogy furnished by experience. It is not, however, impossible for such persons to have the undoubted advantage of a mixed treatment, which combines the utility of altitude and ferruginous waters. In the case of each patient, however, the physician should act as he thinks best. In my opinion, a station situated in the plain should be first recommended, and then, after an interval of fifteen days, during which the altitude becomes gradually higher, St. Moritz may be reached, and the residence there will be not only tolerated but beneficial, while the anæmic condition, of which the pronounced character was at first a counter-indication, will then diminish. Another plan which I adopt is to have the St. Moritz water taken for as long as from four to six weeks in some suitable place, as, for instance, Ragatz* at first, and then to increase the altitude, at first every ten days, and then at shorter intervals. If this plan is adopted, the efficacy of which, in my opinion, is undoubted, the counter-indication mentioned loses its absolute character, being then but a relative and temporary obstacle to this form of treatment.

Long experience has made me familiar with examples of inappropriate treatment in such cases, which may be observed yearly. Who is to be blamed for such mistakes? Rarely,

---

* Ragatz is a town of Switzerland, adjoining the waters of Pfeffers. See "Health Resorts," J. Burney Yeo, M.D.; "The Baths and Wells of Europe," J. Macpherson, M.D.; "Wintering in the Riviera," W. Miller; "Curative Effects of Baths and Waters," Julius Braun; "Switzerland and the Adjacent Parts of Italy," K. Baedeker; "The Central Alps," John Ball; "Handbook to Switzerland" (J. Murray); "Dictionnaire Encyclopédique des Sciences Médicales," Paris; "Hachette's Diamond Guide to Switzerland;" "The Alps and how to see them," J. E. Muddock; "European Guide-book," Appleton; etc.

I regret to say, the patient. A person who has determined to employ treatment at a distance from home would certainly not object to measures which may ensure its success, while the physician too often fails to point out the necessity of these precautions, either from not knowing their importance, or because, his own experience and personal impressions being alone considered, he commits the mistake of supposing that the patient has the same rapid power of adaptation as a healthy person. Owing to observations made during almost twenty years, I have the greatest confidence in the remarkable effects of the climate and waters of St. Moritz,* and it is on this account that I wish patients to benefit from their use, by adjoining to their employment that caution which will enable them to be at once tolerated, and remove any inconvenience which might result from their too sudden action. These principles should be applied with the same strictness when the debility and anæmia do not notably predominate over the nervous and cardiac symptoms. The most that should then be allowed is a diminution of the time to be spent at the intermediate places, though these should still be of the number of two, and at the altitudes already mentioned. The same precautions should naturally be taken when the place is left. The state of the barometer at the different places should be considered, it being remembered that the atmosphere of the plain is condensed when compared with that of the Engadine; this being the case, it would be obviously an error to make a rapid and sudden transition from such a rarefied to a compressed atmosphere. The practical injunctions already given show how this danger may be avoided, and reduce to a minimum degree the counter-indications to residence at St. Moritz.

It might be thought that such care is unnecessary, and that the above precautions are more than is required; it may easily be proved, however, that this is not the case, and that nothing

* See note, p. 250.

unnecessary has been mentioned. Thus, if the counter-indication which results from pronounced anæmia and constitutional debility is unperceived, and the patient, notwithstanding their existence, is sent directly to the Engadine, complications may ensue of a far more serious nature than those which have as yet been mentioned, and which would certainly become alarming should the residence at St. Moritz be blindly continued. It must not be forgotten that, besides the constitutional conditions of this disease, the walls of the blood-vessels are often abnormally liable to rupture, and that residence at this altitude, without any preparation being made, has precisely the same effect as a cupping-glass or pneumatic receiver. Epistaxis may occur daily, and often many times during the day. I have twice known conjunctival hæmorrhage to take place, and three times an eruption of hæmorrhagic purpura over the whole surface, characterized by the eruption itself, the epistaxis, and the ecchymotic marks. The last of these observations was made in the summer of 1880, occurring in a relation of one of my most distinguished colleagues at Paris. The patient was a lady who imprudently directed her own treatment, and went at once to St. Moritz.* Such facts as these, in my opinion, more than justify the rules which I have laid down, both as to the strictness with which indications should be followed, and the precautions which should be adopted with regard to residence in these high regions.

The station of St. Catherine † in the Valteline is less easy to reach from France than that of St. Moritz, though, as far as the mineral contents of its waters and the climatic conditions of the place are concerned, it is in every respect similar to it. Its altitude of 6080 feet (1852 *mètres*) is higher than that of the baths of St. Moritz, and but ten feet (*trois mètres*) less than that of the town. Hence the indications and counter-indications, the precautions which should be taken before and after

* See note, p. 250. † See note, p. 251.

the employment of the mineral waters, are precisely the same. Within the last few years the material conveniences of this place have improved, but the establishment is still small, and the arrangements for bathing insufficient, so that the waters can scarcely be used except as a drink. Were it not for the convenience of being able to live close to the baths, there is, I believe, no circumstance which makes this Italian station more acceptable to the patient than that of the Engadine.

When it is impossible to employ measures which enable patients to derive benefit from the altitude of places which at first sight seem to be indicated, recourse should be had to the ferruginous waters of Schwalbach,* Spa,* or Forges,* stations which are of almost the same value, notwithstanding the difference which their waters present as to richness in iron.

The association of constitutional debility with actual or previous manifestations of scrofula is the characteristic of the third group of cases during the prophylactic period.

If the different pathological connections between scrofula and tuberculosis are not forgotten, the importance of this group will be easily understood. In reality, it is twofold: firstly, on account of the numerical preponderance of these cases; secondly, from the usual efficacy of well-directed treatment. When the manifestations of scrofula are but slight, and its expression of but slight and temporary importance, the prophylactic treatment which has been described, if wholly carried out, will suffice to cure the complaint, and the employment of mineral waters will be unnecessary. Such an event, however, is uncommon, and on this account the indication of mineral waters is almost constant in cases of this nature; on the one hand sulphurous waters, on the other those containing chloride of sodium, or iodine, bromine, and chlorine, are then employed. Which of these is chosen should principally depend upon the form and degree of the scrofulous manifestations which exist.

* See note, p. 251.

The sulphurous waters are of special utility in the superficial, cutaneous, and catarrhal manifestations. The waters which contain sodic chloride, either in large or small amount, are most beneficial when the deeper structures are involved, acting not only upon the visible indications of the disease, but affecting the constitution itself, and lessening the tendency to disease which this engenders. The complex composition of the mineral waters of Uriage * gives special importance to this station, as has been long known. They act in the same beneficial way as the sulphurous waters upon the cutaneous and mucous manifestations of scrofula, and resemble those which contain sodic chloride, in having a pronounced effect upon the constitution. The waters of Aix-la-Chapelle † have also this double composition to a less extent, and in consequence the same twofold therapeutic effect.

The choice of the individual spring among the sulphurous waters should depend principally upon the climatic conditions. It is easy to associate with the action of a mineral water the beneficial effect of altitude and mountain air, an advantage which should not be neglected. The stations in the Pyrenees furnish every desirable variety of such combination—from Luchon,‡ the height of which is 2062 feet (628 *mètres*), to

---

* See note, p. 251.

† Aix-la-Chapelle is in Germany, near the Belgian and Dutch frontiers. See " Curative Effects of Baths and Waters," Julius Braun; " The Baths and Wells of Europe," J. Macpherson, M.D.; " Handbook for Travellers in North Germany " (J. Murray); " Handbook for Travellers on the Rhine," K. Baedeker; " A Three Weeks' Scamper through the Spas of Germany and Belgium," Sir E. Wilson; " European Guide-book," Appleton; etc.

‡ Bagnères-de-Luchon, or Luchon, is a French town to the south of the department Hautes Pyrénées. See " Curative Effects of Baths and Waters," Julius Braun; " Principal Baths of France, Switzerland, and Savoy," E. Lee, M.D.; " The Baths and Wells of Europe," J. Macpherson, M.D.; " Medical Guide to the Mineral Waters of France," etc., A. Vintras, M.D.; " Health Resorts," J. Burney Yeo, M.D.; " Handbook for Travellers in France," part i. (J. Murray); " European Guide-book for English-speaking Travellers," Appleton; " Guide to the South of France," O. B. Black; etc.

Cauterets,* with an altitude of 3257 feet (992 *mètres*) above the level of the sea; in addition to which there is Eaux-bonnes,† the elevation of which is 2593 feet (790 *mètres*). When an inactive form of scrofula exists, with lesions of the bones or joints, the waters of Barèges,‡ at the height of 4200 feet (1280 *mètres*), may be utilized in patients who have reached the age of puberty. At the same time, these waters can rarely be employed on account of the excitement which they produce. This obstacle can, however, be avoided, and the use of sulphurous waters be combined with the benefit of an altitude higher than 3500 feet (1000 *mètres*), by means of recourse to the waters of Gurnigel § in Switzerland, the height of which is 3793 feet (1155 *mètres*). The establishment and arrangements for taking baths are equally good; the altitude is sufficiently high to constitute in itself a means of treatment, and in addition to these advantages the treatment by milk or whey can be employed there without difficulty.

It is scarcely needful to mention the waters of Allevard, ‖ on account of the notoriety which they enjoy. At the same time, on account of their relatively low altitude of 1560 feet (475 *mètres*), little more can be expected of them than the special effects of the mineral waters. They may, however, be employed beneficially in the case of those patients in whom there is so much excitability that residence at a higher altitude than that of 1650 feet (500 *mètres*) is inadvisable. For a similar reason Schinznach ¶ may be recommended, whose altitude of 1250 feet (380 *mètres*) has but little action, while the

---

* See note, p. 24. † See note, p. 278.

‡ Barèges is a French watering-place, department Hautes Pyrénées. See "Dictionary of Watering-Places" (L. Upcott Gill); "The Baths and Wells of Europe," J. Macpherson, M.D.; "Principal Baths of France, Switzerland, and Savoy," E. Lee, M.D.; "The Mineral Waters of France;" etc., A. Vintras, M.D.; "Curative Influence of the Climate of Pau," A. Taylor, M.D.; "Guide to the South of France," C. B. Black; etc.

§ See note, p. 250. ‖ Ibid.

¶ Schinznach is a small Swiss village, in the canton of the Aargau. See

therapeutic effect of the waters upon scrofulous affections of the skin is increased by their combination with the neighbouring waters of Wildegg,* which are remarkable for their richness in chlorides, bromides, and iodides.

When the lesions due to scrofula are more deeply situated, and waters containing sodic chloride are indicated, Saxon should be preferred, a place which will soon be again considered; or La Bourboule,† which, from its waters being rich in arsenic, deserves special consideration, and whose altitude of 2775 feet (854 *mètres*) makes it a powerful agent in restoring strength to the constitution; or, again, Soden,‡ a place signalized long since by Thilenius as efficient in preventing the development of tuberculosis. At the same time, if there are lesions of the bones or joints, indicating the existence of scrofula in a severe form and of old date, the most energetic treatment should be employed. Recourse must be had to the stations at which waters containing a large amount of sodic chloride are associated with saline springs, as at Kreuznach,§ Nauheim,‖ and

"Curative Effects of Baths and Waters," Julius Braun; "The Baths and Wells of Europe," J. Macpherson, M.D.; "Guide to Switzerland," K. Baedeker; "Handbook for Travellers in Switzerland" (J. Murray); "Dictionnaire Encyclopédique des Sciences Médicales," Paris; etc.

* Wildegg is also a watering-place in Switzerland, between Bâle and Zurich. See "Curative Effects of Baths and Waters," Julius Braun; "Guide to Switzerland," K. Baedeker; "Hachette's Diamond Guide to Switzerland;" etc.

† See note, p. 246.

‡ Soden is a German village, immediately beneath the ancient town of Königstein. See "Dictionary of Watering-Places," part ii. (L. Upcott Gill); "Health Resorts," J. Burney Yeo, M.D.; "Curative Effects of Baths and Waters," Julius Braun; "Handbook for Travellers in North Germany and on the Rhine" (J. Murray); "Handbook for Travellers on the Rhine," K. Baedeker; etc.

§ Kreuznach is situated in the valley of the Nahe, in Germany. See "Dictionary of Watering-Places," part ii. (L. Upcott Gill); "Health Resorts," J. Burney Yeo, M.D.; "Curative Effects of Baths and Waters," Julius Braun; "Handbook for Travellers in North Germany and on the Rhine" (J. Murray); "Handbook for Travellers on the Rhine," K. Baedeker; etc.

‖ Nauheim is a small German town, in the Wetterau valley. See "Curative Effects of Baths and Waters," Julius Braun; "Health Resorts," J. Burney Yeo, M.D.; "Handbook for Travellers in North Germany and on the Rhine" (J. Murray); "Handbook for Travellers on the Rhine," K. Baedeker; "European Guide-book for English-speaking Travellers," Appleton; etc.

Salins* in the Jura. The mixed composition of the waters of Uriage† enables them to be also employed in such circumstances.

The most important of all scrofulous symptoms in this disease is, without doubt, persistent enlargement of the lymphatic glands, whether superficially or deeply situated. This lesion has already been shown to have the power of producing tubercle (*phymatogène*). Whilst it exists, the patient is in twofold danger of tuberculosis: firstly, on account of the constitution and scrofulous diathesis; secondly, from the mere existence of enlarged glands, which may become at any time the foci or agents of secondary infection in a more or less generalized form. Such a condition requires every care on the part of the physician, and when the therapeutic treatment of scrofula with respect to the prevention of phthisis is considered, this symptom has a most serious interest; the endeavour to remove any such disorder should be ceaseless, and this task will not be accomplished until the glandular enlargement has entirely ceased to exist. Without doubt, all the mineral waters named, and notably those of La Bourboule,‡ Soden,§ Kreuznach,∥ Salins, to which those of Reichenhall ¶ may be added, are of real benefit in this disorder; but, according to my experience, those of Saxon,** in Switzerland, hold the first place in importance, with an altitude of 1750 feet (534

* Salins is a French town, in the department of the Jura. See "Medical Guide to the Mineral Waters of France," A. Vintras, M.D.; "Handbook for Travellers in Switzerland," K. Baedeker; "Handbook for Travellers in France" (J. Murray); "Dictionnaire Encyclopédique des Sciences Médicales," Paris; "Hachette's Diamond Guide to Switzerland," etc.

† See note, p. 251.   ‡ See note, p. 246.
§ See note, p. 261.   ∥ Ibid.

¶ Reichenhall is a German town near Salzburg, in Upper Bavaria. See "Curative Effects of Baths and Waters," Julius Braun; "Handbook for Travellers in South Germany and Austria" (J. Murray); "Eastern Alps," John Ball; "Handbook for Travellers in the Alps," K. Baedeker; etc.

** Saxon is in the Martigny district of Switzerland, in the canton of Valais. See "The Baths and Wells of Europe," J. Macpherson, M.D.; "Guide to Switzerland," K. Baedeker; "Handbook for Travellers in Switzerland" (J. Murray); "Hachette's Diamond Guide to Switzerland;" "European Guidebook for English-speaking Travellers," Appleton; etc.

mètres). These waters contain calcic and magnesic bicarbonate, with iodine and bromine, and have been employed by me during the last few years with remarkable success, not only in cases of superficial glandular enlargement, but also, in 1879, in a case of considerable infiltration of the bronchial glands. Owing to their composition, and specially their richness in potassic iodide, the mineral waters of Challes,* in Savoy, have in all probability the same effect, but with regard to this fact I have not had the same experience.

Such are the principles which guide me in the employment of mineral waters during the premonitory period of tuberculosis. Their employment is, in my opinion, of great importance to those who cannot have the advantage of complete hygienic and climatic treatment, and it is during this time that mineral waters have, as I think, their widest sphere of action and greatest effect.

The confirmed disease in its ordinary form will now be considered.

Before proceeding further, it should be said that the reserve already made with respect to the preponderating importance of prolonged climatic treatment must not be forgotten. Whenever the condition of the patient and of the disease enables this to be put into practice, whenever its adoption confirms from the first the propriety of the indication, nothing more need be done. All possible benefit is obtained by such means, and care should be taken not to compromise its action by employing mineral waters, of which the most certain effect is to cause abandonment of more effective treatment. Unfortunately, as has been already stated, the climatic treatment can but rarely be employed in practice, not from absence

* Challes is situated in Savoy, at a distance of about three miles from Chambéry. See "Medical Guide to the Mineral Waters of France," etc., A. Vintras, M.D.; "Handbook for Travellers in Switzerland" (J. Murray); "Handbook for Travellers in Switzerland," K. Baedeker; "Dictionnaire Encyclopédique des Sciences Médicales," Paris; etc.

of indication, but because the patients in most cases cannot or will not submit to it. At the same time, though it can but rarely be employed, the importance of this remedy should be borne in mind, and it is much to be regretted that it can so rarely be adopted, full liberty of choice being alone allowed with regard to the use of mineral waters.

Like other modes of treatment, mineral waters are the more useful in proportion as they are employed at a time nearer to the onset of the disease, and therefore when the lesions are less pronounced and the constitution less affected. At the same time, it would be a mistake to consider the degree of pulmonary lesion as the fundamental criterion whether mineral waters should be employed or not. Even when caverns exist, the waters may be beneficial, should there be but one or few cavities, of small size and surrounded by healthy tissue; the size of the lesions is more important than their degree, and this alone, without other consideration, may prevent the employment of mineral waters. The leading counter-indication, however, and the most important, since it may occur at any period of the disease, is the existence of pyrexia.

Whatever the condition of the lesion and the general state of the patient, whatever the origin of the fever may be, whilst pyrexia exists the patient should on no account be treated by mineral waters. As regards inflammatory fever, that due to the formation of caverns, absorption, or the development of tubercles combined with secondary granular formations, this counter-indication is absolute. With regard to the fever which attends the first development of tubercles, whether of remittent, tertian, or quotidian type, the same may be said. This indication should be strictly followed unless, as in many cases, the fever is of no regular type, and only exists temporarily at more or less distant periods. If all the other circumstances of the case indicate the employment of mineral waters, the existence of this symptom may in that case be disregarded from the fact that the improvement produced thereby in the general condition and local lesion may cause the vague febrile symptoms to pass away. The choice of the watering-place must be then made with the greatest care. The counter-indication due to pyrexia has no indefinite character, continuing during the whole time of the fever, and even for a mean period of three months after its cessation. If after this time fever is still absent, and the condition of the patient is not incompatible with the change of residence and fatigue associated with the employment of mineral waters, the special obstacle due to the existence of fever will, in my opinion, have ceased to exist, and the physician will be justified in employing whichever plan of treatment seems most beneficial to the patient. The choice of the mineral water will be then an important question, and a mistaken selection may have unfortunate consequences.

The existence of actual or recent hæmoptysis is, it need scarcely be said, an equally imperious counter-indication. Of hæmoptysis which took place at an old date the same cannot be said, even though it may have occurred repeatedly, and been associated with fever, showing it to be connected with the existence of congestion. Mineral waters are not excluded from the treatment by this fact alone, so long as during several months the patient has been free from pyrexia or hæmoptysis. On the other hand, if a judicious selection is made, they may then act powerfully in preventing the recurrence of these complications. Lastly, whether it occurs early or late in the disease, wasting definitely excludes their employment, and non-compliance with this rule is the cause of numerous and sad disappointments.

These fundamental precepts have been only mentioned on account of the condensed nature of this review; but they should be understood as applicable in every form, at every period, and in every case of the disease.

The elements which determine the choice of the mineral

water, or, in other words, the sources of the indications according to which it is made, are two in number: firstly, the primary or secondary nature of the tuberculosis; secondly, the power of reaction on the part of the patient. The first, which also precedes the other chronologically, should be specially investigated by the physician, who must carefully settle whether the pulmonary disease, arising spontaneously and independently of any other pathological condition, should be looked upon as due to the tubercular diathesis and general debility which are the common source of every form of tuberculosis, in which case the phthisis is primary, or, on the other hand, coexists with actual or previous constitutional disease, to the existence of which it may be supposed due, when the phthisis would be secondary, that is, scrofulous, arthritic, or herpetic.

This point having been decided, the second question becomes of preponderating importance. Until it is solved, the class of mineral waters to be employed must remain unsettled, since, though each of the three groups of secondary phthisis may, perhaps, be looked upon as undoubtedly answering to one of the groups of mineral waters, this is not the case in primary phthisis, which itself indicates the employment of no special mineral water, and cannot be looked upon as a guide or indication. The selection of an individual place in the group, which is of true practical interest, is entirely dependent upon the second question at issue, namely, the power of reaction on the part of the patient. Whether the phthisis be primary or secondary, and if secondary whether it be scrofulous or arthritic, is of little consequence, since a good selection can only be made by taking as guide the relation which exists between the patient and the disorder—that is to say, the manner in which the organism tolerates the influence of the existing disease. This, which is of purely individual character, can only be ascertained by means of medical knowledge, and the difference which exists in individual patients, as already said, produces an insuperable obstacle to therapeutic dogmatism.

Some general facts, however, may be mentioned which are true in every case.

The relation between the patient and his disease, depending upon the individual power of reaction, presents two types which approach each other by numerous intermediate characters, but which are always to be distinguished. In one of these the patient is extremely sensitive to the influence of the disease, and the reaction is proportional to, or perhaps exceeds, its real strength. Bearing but impatiently the effects of the complaint, the organism betrays its condition by constant excitability of the heart and nervous system, the perpetual liability to fever which but too often exists, the frequence and intensity of acute disorders of the respiratory organs, perhaps hæmorrhagic in character, these being symptoms which but too often distinguish what may be termed the active or florid type of organic reaction. The other type, even when most pronounced, is distinguished by a condition which nothing seems to affect; the lesions develop, harmful symptoms multiply, the strokes of the disease become more intense, and fever may develop, while the organism, owing to its innate torpidity, seems to remain unmoved. It tolerates and seems disposed to tolerate everything without reaction, or if perchance it reacts, does so but feebly, and as if involuntarily for a time, after which it returns to the same languid condition. This absence of excitability, the rare occurrence of but slight pyrexia occurring late in the disease, and silent extension of the lesions without manifest symptoms, are features which characterize the passive or torpid type of organic reaction.

The expressions florid or active, and torpid or passive phthisis have often been criticized, and without doubt are not properly applicable to the disease itself. It is not the complaint which is florid or torpid, but the patient, who, on account of his nature, his physical and moral individuality, reacts in an active or torpid manner, and thus imprints a special character on the disease which affects him. The above denominations should, therefore, be looked upon merely as brief expressions, which indicate at once and without need of comment the reacting power of the patient. As thus understood they should be preserved, being convenient, and containing an idea which guides both the prognosis and treatment. It is now intelligible how important these facts are as regards the effect of watering-places, and that it is impossible, without danger of harming the patient, to do otherwise than base the treatment upon his individual character. The rule to be taken as guide is difficult and at times uncertain, but the theoretical principle upon which it is based is both clear and precise. Thus an active or florid temperament, however it is called, counter-indicates the use of mineral waters, which have a stimulant effect, while a torpid one requires their employment. As a general principle this is an indisputable fact, the twofold character of the reaction not only determining the question of treatment by mineral waters, but deciding which class of waters is indicated by the cause and progress of the disease. In applying, however, these principles difficulties still exist, from the existence of varieties in the type of reaction of such infinite number as to have an individual character. This difficulty would at any rate partly cease, if the waters which are appropriate to the treatment of phthisis could be divided into two classes, comprising respectively those of which the effect is stimulant or sedative. Such a division, however, is impossible, since each of the mineral waters may produce excitement in a given organism. With respect to this point, there is in fact but a question of degree, so that in practice the principle already stated may be more truly and beneficially expressed in the following terms. The florid or active type of reaction indicates employment of the least, the torpid or passive type of the most stimulant mineral waters. In my opinion, it is impossible to say more than this with regard to application of the general rule, and selection of the individual place must be dependent upon the individual concerned.

This selection, again, by no means entirely depends upon the two fundamental facts just mentioned, namely, the character of the patient on the one hand, and the stimulant effect of the mineral water on the other. The climatic conditions of the place, which are so often neglected, must be taken into consideration, being of no less importance than the composition of the water. These may be most appropriate in the case considered, while the other particulars of the place counterindicate its use. The altitude, with its exciting effect, which increases proportionately to the height of the place, should be specially borne in mind; as also its exposed condition, the frequence, direction, and intensity of the winds, the humidity of the atmosphere, and, in fact, all the meteorological conditions of the locality. Nor is this all; the condition of the gastro-intestinal functions of the patient, and the probable effect of the waters upon the digestive organs, must be considered, while enquiry should be also made as to the sanitary condition of the place, not only at ordinary times, but also as regards modifications which are liable to occur from the effect of the season. Thus it is not extremely rare, in mountainous regions, for intestinal catarrh to prevail epidemically in the summer, with or without dysenteric symptoms. While such an accidental condition exists the station should on no account be employed, since the effect of this disorder upon a patient suffering from tuberculosis might be most injurious, continuing, perhaps, when the place itself is left.

There are yet other details which might be mentioned in connection with this point, but the fundamental principles according to which the decision should be made have now been stated, and it has been fully shown how important, multiple, and diverse are the elements of this therapeutic decision, which cannot be disregarded without danger to the patient.

It will now be briefly explained how the general principles already mentioned can be applied in practice.

Primary Phthisis.—In this case, no constitutional disease with special indications having previously existed, the action of the mineral water, however decided it may be, merely affects the general nutrition and the peritubercular lesions of the bronchial tubes, having no effect, as already stated, upon the tubercles themselves. This twofold effect may be produced, either by mineral waters containing sodic chloride in a simple or complex form, or by chlorides associated with bicarbonates, calcic or sodic sulphates, or sulphurous acid, all of which may be of real benefit. In order to take the right direction in such a labyrinth of mineral wealth, the power of reaction in the patient must be specially considered, that is to say, the ordinary brief expression being used, whether the phthisis is of the florid or torpid type, and whether acute phases have or have not existed in the previous evolution of the disease; it is only then that, in order to utilize the notions contributed by the patient, the degree of stimulation produced by the mineral water and by the climate of the different stations should be understood.

In my opinion, a classification formed upon this base can alone be of any practical utility. The chemical action of the water may, perhaps, seem to be disregarded; but the duty of the physician is undoubtedly to give the first place to whatever is of most importance in guiding his decision. Thus all the mineral waters which might be beneficial in the treatment of primary phthisis may be divided into three groups, the effect both of the waters and climate, specially as regards the altitude of the place, being taken into consideration. Long experience has continued to show the advantage and justness of this new method of arranging the places, which is the following:—

1. Mineral waters and stations with slightly stimulating effect.
2.　　　　„　　　　„　　　　　　moderately　　„
3.　　　　„　　　　„　　　　　　decidedly　　　„

The first group would be specially employed in cases of pronounced morbid excitability, of hæmoptysis, and when the disease has been accompanied by acute complications.

The second group is adapted to patients in whom the disease, without belonging actually to the torpid type, properly so called, is accompanied by a reaction of less intensity than in the preceding cases. Hæmoptysis is absent, or, if it occurs, is but slightly pronounced, and unassociated with fever, while the progress of the tuberculosis has been free from acute complications of any notable character. In both these groups, specially the first, temporary attacks of intermittent fever, due to the development of tubercles, should not, as already explained, be looked upon as an absolute counter-indication.

Lastly, the third group is adapted to patients in whom the reaction is torpid; the absence of hæmoptysis, of pneumonia, or broncho-pneumonia combined with fever, or of actual pyrexia in the antecedent history, being, in my opinion, indispensable conditions for the employment of mineral waters. On account of their pronounced effects, the mineral waters of this group should not be used unless the conditions plainly indicate their use, or if any doubt exists as to whether this or the previous group should be employed. In such a case it would be more prudent to administer the mineral waters belonging to the second group.

The individual places in each group have not a similar effect, and present numerous varieties in the degree of stimulation which they produce; in the enumeration of stations which will shortly be given the names of the mineral waters which are least active in this respect, either in themselves or owing to their climate, will be first mentioned. The places differ also to a notable extent with respect to therapeutic and other requirements, details which cannot be considered in this work, though the physician should be informed upon all these points, that, while having the best intentions possible, he may not be liable to inflict injury both upon the patient and himself.

First Group.—This group is adapted to the most pronounced degree of morbid excitability, to those who suffer from hæmoptysis, or in whom the disease has been associated with acute complications.

Ems * and Soden † are the types of this group. The altitude of both these places is insignificant, that of the first being but 328 (100 *mètres*), and that of the second 484 feet (145 *mètres*), above the level of the sea. The mineral waters of Ems, which contain chlorinated bicarbonate of soda, and of Soden with chlorinated soda, owe to these circumstances their efficacy, not only over the catarrhal conditions, but also over the nervous and cardiac excitability of the patients who employ them. Soden is also said to remove the tendency to hæmoptysis existing in those who suffer therefrom, a fact which is true though not constant. Thus a patient under my care, subject to frequent attacks of congestion, attended by hæmoptysis, was affected by this complication towards the end of his treatment by residence at that place. At the same time, while an absolute cure cannot be expected, these waters are the best which can be employed by those suffering from this tendency.

Ischl ‡ and Royat, § whose stimulating influence is slightly

---

* Ems is a German town, on the right bank of the Lahn. See "Curative Effects of Baths and Waters," Julius Braun; "Health Resorts," J. Burney Yeo, M.D.; "Handbook to North Germany and the Rhine" (J. Murray); "Handbook for Travellers on the Rhine," K. Baedeker; "Handbook for Travellers in Switzerland" (J. Murray); "Dictionnaire Encyclopédique des Sciences Médicales," Paris; "A Three Weeks' Scamper through the Spas of Germany and Belgium," Sir E. Wilson; "European Guide-book," Appleton; etc.

† See note, p. 261.

‡ Ischl is an Austrian town, in the mountainous region of the Salzkammergut. See "Curative Effects of Baths and Waters," Julius Braun; "Handbook for Travellers in South Germany and Austria" (J. Murray); "Eastern Alps," John Ball; "Handbook for Travellers in the Eastern Alps," K. Baedeker; "European Guide-book," Appleton; etc.

§ Royat is a French town, in the department of Puy-de-Dome. See

more pronounced, on account of their altitude rather than the mineral water, may, in my opinion, be ranged in the same group. The altitude of Ischl, in the Salzkammergut of Austria, is 1410 feet (430 *mètres*). Surrounded by pine-forests, it has also the advantages of a mild subalpine climate, and its waters, containing chlorinated soda, which are usually taken in combination with whey, have without doubt a beneficial effect upon the catarrhal complications of tuberculosis, specially at its onset. This fact has been proved again and again. In the same circumstances recourse may be had to the waters of Royat in Puy-de-Dome. Their composition resembles that of the waters at Ems, from which they differ in containing more iron. Their altitude of 1490 feet (450 *mètres*), however, which produces a climatic effect, will not allow them to be classed in the same group.

Considered with respect to their local action, the above-mentioned mineral waters have most effect upon those cases in which the peritubercular lesions are catarrhal in nature. These are only found at the onset of the disease, and when the lesions consist of compact infiltrations or indurations due to pneumonia or broncho-pneumonia, the power of these waters upon the progress of the disease is less, if it exists at all, and recourse should be then had to two other places, whose character enables them to be placed in the first group, namely, Lippspringe * and Weissenburg.† The springs of both these places contain calcic sulphate, while the slight altitude of

---

"Curative Effects of Baths and Waters," Julius Braun; "Health Resorts," J. Burney Yeo, M.D.; "Medical Guide to the Mineral Waters of France," A. Vintras, M.D.; "Handbook for Travellers in France" (J. Murray); "Dictionnaire Encyclopédique des Sciences Médicales," Paris; "Auvergne, its Thermo-mineral Springs, Climate," etc., R. Cross, M.D.; "Guide to the South of France," C. B. Black; etc.

* Lippspringe is a German station in Westphalia. See "Handbook for Germany and the Rhine" (J. Murray); "Dictionnaire Encyclopédique des Sciences Médicales," Paris; etc.

† See note, p. 274.

Lippspringe, in Westphalia, causes it to have the same action as the climate of plains. The mean hygrometric condition of the place is also so great that this element forms one of the therapeutic agents at a station which is specially adapted to diseases in which excitement is apt to occur. Weissenburg,[*] in Switzerland, in the canton of Berne, is placed at the altitude of 2920 feet (890 *mètres*). Besides the sedative and alterative action of the mineral water, there is also the slightly exciting effect of a mountain climate, which is tonic rather than stimulant. Remembering these facts, one or other of the watering-places should be chosen according to the excitability and constitutional state of the patient. The mineral waters of both are well adapted to the cases which are now being considered, conferring much benefit on those whose constitution is not shattered by the disease, who are not too much weakened by its effects, and who have more than once suffered either from attacks of congestion with hæmoptysis, or of broncho-pneumonia which have left persistent lesions in the lung. There is, in my opinion, no reason why the employment of these two places has been, and perhaps still is, reserved until the advanced period of phthisis. Such is a complete mistake, since, as with other modes of treatment, their effect is the more pronounced in proportion as they are employed at an earlier time in the complaint, when they may be of undoubted benefit. In accordance with this principle, they are both useful in the catarrhal complications of the disease which were first mentioned, though in my opinion, unless the special conditions of the patient counter-indicate its use, Weissenburg should be preferred on account of its altitude.

Second Group.—This group, as already stated, is adapted

[*] Weissenburg is a Swiss town, in the canton of Berne, near Thune. See "Curative Effects of Baths and Waters," Julius Braun; "The Baths and Wells of Europe," J. Macpherson, M.D.; "Handbook for Travellers on the Rhine," K. Baedeker; "Handbook for Travellers in Switzerland" (J. Murray); "The Alpine Guide," John Ball; etc.

to patients in whom the illness, without being of the torpid form, properly so called, presents a reaction, an excitability which is less pronounced than in those of the first class, who have not suffered from hæmoptysis at all, or only to a slight degree, without the association of fever, and in whom the course of the tuberculosis has from the first been free from acute manifestations.

The group should be composed, in my opinion, of certain stations at which the waters contain sulphurous acid, soda, and lime, of one spring without analogue which may be termed nitro-saline, and of two which contain arsenic.

Amongst the sulphurous waters I would first mention, on account of the small degree of excitement which they produce, the stations of St. Honoré * in the Nièvre, and Amelie-les-Bains† in the Eastern Pyrenees. Both have an altitude of about 900 feet (275 mètres), which would be incompatible with any stimulant action of climatic origin; the waters at both stations contain sulphurous acid in small quantity, representing the weakest sulphurous springs which are employed medicinally. These are, in fact, the least exciting waters of the second group.

Vernet,‡ in the neighbourhood of Amelie, and Allevard,§ in the Isère, belong to the same group. As regards the former, the difference in the degree of stimulation results from its altitude, which slightly exceeds 1970 feet (600 mètres), while in the latter, an altitude of 1560 feet (475 mètres) is combined with the special action of the mineral water in producing this

---

* St. Honoré is a French town in Nièvre. See "Medical Guide to the Mineral Waters of France," A. Vintras, M.D.; "The Baths and Wells of Europe," J. Macpherson, M.D.; etc.

† See note, p. 246.

‡ Vernet is a place in France, in the department of the Eastern Pyrenees, at the foot of the Canigou. See "Curative Effects of Baths and Waters," Julius Braun; "Medical Guide to the Mineral Waters of France," A. Vintras, M.D.; "Handbook for Travellers in France" (J. Murray); "Guide to the South of France," C. B. Black; etc.

§ See note, p. 250.

effect. These facts being remembered, both may be employed with benefit by patients to whom the second group of waters is specially adapted.

The station of Panticosa,* on the Spanish side of the Pyrenees, should be classed with these places. Owing to its altitude, which exceeds 6560 feet (2000 *mètres*) (Seco-Baldor), this station is undoubtedly the most stimulating, on account of its climatic effect. On the other hand, however, its mineral waters have special sedative properties, which to some extent compensate for the exciting effect of its altitude, so that if the patient presents none of the symptoms already mentioned with reference to St. Moritz, as counter-indicating the employment of a high altitude, the waters of Panticosa may be safely recommended. Neither the previous occurrence of hæmoptysis in the conditions already mentioned, nor the stage of the disease, is a counter-indication. The Duke of R——, whose wonderful recovery has already been mentioned, visited Panticosa when a cavern existed in the lung. Hæmoptysis had occurred more than once, while the natural indifference of the organism had been but little disturbed. The fact is that this, like other questions, must be decided by the constitutional condition and reacting power of the patient. The waters are usually classed in France among those containing sodic sulphate, a designation which omits the character that specially characterizes them, namely, that of containing a large quantity of nitrogen; thus the qualification of being saline and containing nitrogen, given to it in Spain, is far more appropriate. The station, easily reached from the Spanish side, is less accessible from France, a reason perhaps for its waters being so little known in the latter country—a fact to be regretted, since they are beneficial in certain definite conditions, and cannot be replaced by the mineral waters of France in the neighbouring districts. These

* See note, p. 24.

are certainly less exciting on account of the climate, but the action of the waters themselves is more energetic, so that they would not be adapted to patients who are excitable, of a somewhat active disposition, or in whom hæmoptysis has occurred, while on the other hand the waters of Panticosa could be employed by them with advantage and without danger.

The composition of this group will be completed by considering the waters of La Bourboule,* which contain sodic chloride and arsenic, and of Mont Doré,† which contain arseniated bicarbonates. In secondary phthisis they are adapted to a special complication, which will soon be reconsidered. In primary phthisis these stations, on account of the excitement produced, have an intermediate character, being in my opinion more exciting than the other places in the second group, but less so than those in the third. The excitement produced is principally due to the action of the water, but also to their altitude, which, in the case of La Bourboule, is 2790 feet (850 *mètres*), and in that of Mont Doré 3450 feet (1050 *mètres*). The forms of disease to which these waters are adapted are not, again, those which, by their torpidity, characterize the third group, though notably resembling them. In my opinion, they should only be employed when any sign of excitement which may previously have existed has entirely disappeared; when there is no pyrexia, when hæmoptysis has never occurred, or at any rate not for several months; when such complications as pneumonia or broncho-pneumonia, if they have occurred at all, have for a long time and totally lost their acute character, and when no form of enteritis either

---

* See note, p. 246.

† Mont Doré is a small valley in the department Puy-de-Dome, France. See " Dictionary of Watering-Places," part ii. (L. Upcott Gill) ; "The Principal Baths of France, Switzerland, and Savoy," E. Lee, M.D. ; ".Medical Guide to the Mineral Waters of France," etc., A. Vintras, M.D. ; " Health Resorts," J. Burney Yeo, M.D. ; " Handbook for Travellers in France" (J. Murray) ; "Auvergne, its Thermo-mineral Springs, Climate," etc., R. Cross, M.D.; " Guide to the South of France," C. B. Black; etc.

exists or has existed, this being an absolute counter-indication to their use. In making a choice between these two places, much importance should be attached to the difference in their altitude and other climatic conditions, though, as will be seen with regard to secondary phthisis, it is the constitutional condition which is of special importance in making the selection.

Third Group.—As already stated, this group of waters is adapted to patients in whom little or no reaction occurs. The non-occurrence of previous hæmoptysis, the absence of pneumonia or broncho-pneumonia complicated by fever, or of pyrexia due to the disease, are, as I believe, indispensable conditions for the employment of these waters.

I look upon this group as composed of three stations in the Pyrenees, namely, in alphabetical order, Bagnères-de-Luchon,* Cauterets,† and Eaux-bonnes.‡ At each of these stations the water of the different springs varies, not only in temperature, but also in richness of mineral contents, so that in employing them the individual excitability should be considered, which, even in the most inactive, presents numerous shades of difference. Bagnères-de-Luchon, with an altitude of 2060 feet (628 *mètres*), is the least highly situated of the three stations, and since the altitude of Eaux-bonnes is 2590 (790 *mètres*), and of Cauterets 3250 feet (990 *mètres*)—a difference which should not be forgotten—it is possible to vary the degree of climatic action. Enteritis in any form counter-indicates the employment of these waters; besides

---

* See note, p. 259. † See note, p. 24.

‡ Eaux-bonnes is a small French hamlet in the arrondissement of Oloron, in the Basses Pyrénées. See "Dictionary of Watering-Places," part ii. (L. Upcott Gill); "Curative Effects of Baths and Waters," Julius Braun; "Principal Baths of France," etc., E. Lee, M.D.; "The Baths and Wells of Europe," J. Macpherson, M.D.; "Mineral Waters of France," etc., A. Vintras, M.D.; "Health Resorts," J. Burney Yeo, M.D.; "Handbook for Travellers in France" (J. Murray); "Curative Influence of the Climate of Pau," A. Taylor; "European Guide-book," Appleton; "Guide to the South of France," C. B. Black; etc.

which the sanitary condition of the place at the time of occupation should be ascertained in each year, since the effect of summer may be to institute a tendency to intestinal catarrh. When this occurs the place should not be visited, that the effect of the waters may not be compromised.

Secondary Phthisis.—The fundamental indication is furnished in this case by a constitutional disease, to the existence of which the development of phthisis may be attributed. The first place is held by scrofula, not only on account of its frequence, but also from the certainty of its causal connection, which is but a possibility in the case of arthritis or herpes. The two latter, indeed, can but rarely be considered as the cause.

In strumous phthisis the most beneficial watering-places are those already mentioned when considering the prophylaxis of the disease in persons affected by scrofula. The selection will depend upon the nature of the scrofulous symptoms, which have been or are still present, as already explained, and upon the power of reaction existing in the patient, which indicates the treatment of the disease when combined with more or less excitement. The preceding classification contains the data necessary for the solution of this question, and a fuller explanation is not needed. It should merely be mentioned that the station of Saxon * is specially adapted to cases of tuberculosis in which glandular swellings exist, while on the other hand La Bourboule,† Uriage,‡ the waters of Gurnigel,§ and stations in the Pyrenees,‖ which contain sulphurous acid, are appropriate to the scrofulous form of phthisis, of which torpidity is so frequent a characteristic.

When tuberculosis affects a person in whom or in whose

---

* See note, p. 262. † See note, p. 246.
‡ See note, p. 251. § See note, p. 250.
‖ Stations in the Pyrenees containing sulphurous acid are Amelie-les-Bains, Allevard, Panticosa, etc.

family gout or rheumatism have undoubtedly existed, when the development of pulmonary complications closely follows the occurrence of the usual gouty or rheumatic manifestations, the question of arthritic phthisis may be taken into consideration. If the climatic conditions seem in other respects adapted to the state of the patient, no more appropriate watering-place could be then chosen than Mont Doré,* Baden † in Switzerland, or Bath ‡ in England, the waters of which contain calcic sulphate. At the same time, the two latter places are only appropriate when the lesion of the joints is still in an active condition. A reserve, however, which was previously stated in my Clinical Lectures, should be here repeated: the relation as to cause and effect between this constitutional disease and the lesion in the chest is always far more uncertain than in cases of scrofula, and consequently, should the employment of mineral waters fail to show by the rapid improvement produced that a real connection exists between the two complaints, it is better not to continue this plan of treatment, but to consider the existence of gout or rheumatism as a simple coincidence, and that in reality the phthisis is a primary disease.

The existence of herpetic phthisis is even more doubtful, and I do not remember to have ever seen an undoubted case

---

* See note, p. 277.

† Baden is a Swiss village in the canton of the Aargau, at the distance of thirteen miles from Zurich. See "Dictionary of Watering-Places," part ii. (L. Upcott Gill); "Curative Effects of Baths and Waters," Julius Braun; "Baths and Wells of Europe," J. Macpherson, M.D.; "Guide to Travellers in Switzerland," K. Baedeker; "Handbook for Travellers in Switzerland" (J. Murray); "The Alps and how to see them," J. E. Muddock; "Hachette's Diamond Guide to Switzerland;" "European Guide-book," Appleton; etc.

‡ Bath is the well-known town at the south-eastern extremity of the county of Somersetshire. See "Dictionary of Watering-Places," part ii. (L. Upcott Gill); "Curative Effects of Baths and Waters," Julius Braun; "The Baths and Wells of Europe," J. Macpherson, M.D.; "Handbook for Travellers in Wiltshire, Dorsetshire, and Somersetshire" (J. Murray); "The Bath Thermal Waters," J. K. Spender; "European Guide-book," Appleton; "Guide to the Roman Baths at Bath," C. E. Davis; etc.

of this disease. At the same time, there are good reasons for admitting a relation between herpetism and tuberculosis, and the indication should not be disregarded. As to treatment, the waters containing sulphurous acid, notably those of Schinznach,* or of Loëche † in Switzerland, a place with regard to which the altitude of 4640 feet (1415 *mètres*) should be specially borne in mind, of St. Gervais ‡ and Uriage, § which contain chlorine, sodium, and sulphurous acid, and of La Bourboule, ‖ containing chlorine, sodium, and arsenic, might be utilized.

Pneumonic Phthisis.—This form of tuberculosis can but rarely be treated by means of mineral waters, such an opportunity only occurring when the disease has passed into a chronic state. It is only when this condition persists without improvement during many months that the mineral waters may be taken into consideration, should the patient's condition seem to suggest their employment. On account of the special mode of onset, and the extremely and rapidly serious consequences which any very stimulating form of treatment might have, the choice is but limited, and should, in my opinion, be confined to one of the stations belonging to the first group, namely, Lippspringe.¶ It is but rarely, however,

---

* See note, p. 260.

† Loëche-les-Bains, Leuk, or Loëche, is a place in Switzerland, Canton Valais, at the junction of the Rhone and Dala. See "Curative Effects of Baths and Waters," Julius Braun; "Dictionary of Watering-Places," part ii. (L. Upcott Gill); "The Baths and Wells of Europe," J. Macpherson, M.D.; "Handbook for Travellers in Switzerland" (J. Murray); "Handbook for Travellers in Switzerland," K. Baedeker; "Dictionnaire Encyclopédique des Sciences Médicales," Paris; "Hachette's Diamond Guide to Switzerland;" "The J.E.M. Guide to Switzerland," J. E. Muddock; etc.

‡ St. Gervais is a much-frequented station in Savoy, at the foot of Mont Blanc. See "Guide to the Western Alps," John Ball, F.A.S.; "Handbook for Travellers in Switzerland" (J. Murray); "Dictionnaire Encyclopédique des Sciences Médicales," Paris; "Principal Baths of France, Switzerland, and Savoy," E. Lee, M.D.; "Baths and Wells of Europe," J. Macpherson, M.D.; "Handbook for Travellers in Switzerland," K. Baedeker; "Hachette's Diamond Guide to Switzerland," A. and P. Joanne; etc.

§ See note, p. 251.   ‖ See note, p. 246.   ¶ See note, p. 273.

that mineral waters would be employed in these circumstances, experience showing what inconveniences may occur, and the superiority of climatic treatment.

A last question presents itself for solution, of general character, and which should not be disregarded. The treatment of pulmonary phthisis, as already said, comprises four groups of agents, namely, that which is hygienic, including hydro- and aërotherapeutics, medicinal remedies, mineral waters, and climate. A question may, therefore, be raised as to the position which should be assigned to mineral waters as compared with these distinct therapeutic agents. The answer depends upon whether the application of other forms of treatment is fully adopted. If the hygienic and medicinal remedies are employed strictly according to the rules previously mentioned, if the climatic treatment is carried out in conformity with principles which will soon be stated, the answer may be made without hesitation or doubt, that in these conditions mineral waters must take the last place in the series. Only beneficial when patients are incompletely treated by other means, they cannot be looked upon as a fundamental or necessary means of treatment.

CHAPTER XI.

CLIMATIC TREATMENT.

Therapeutic uniformity with regard to the effect of climate in pulmonary phthisis—Causes of this error—Difference in the indications of climate as in those of other therapeutic agents—Elements of appreciation—Preliminary questions to be solved.

Immunity of certain climates from phthisis—Conditions of this immunity—Influence of altitude, mineral waters, and mode of life upon the liability.

Effects of high climate upon the organism—General effects upon the nutrition and strength—Special effects upon the respiratory functions—Relations between the climatic effects, and the fundamental indications of the treatment of phthisis.

New division of climates—Respective therapeutic action—Indications and counter-indications—Conclusion.

It is a fact, in my opinion to be regretted, that the influence of long-existing routine, with but few exceptions, guides the medical practitioner when he selects the most appropriate climate in the treatment of pulmonary phthisis. From the existence of some impression believed rather than proved to be correct, the idea of this disease is intimately associated as regards treatment with that of warm climates. Thus in the case of patients affected thereby, one may be suffering from confirmed phthisis, another from an early stage of the malady, while a third has not yet the slightest trace of tuberculosis, though for some reason the subsequent development of the disease is to be feared. These three categories, which include all cases, both of phthisis and of those who are liable to the complaint, are so dissimilar, and require such different hygienic and pharmaceutical remedies, that no uniform plan of treatment can possibly be admitted as correct. Still, when the question of climatic treatment is considered, this difference is either overlooked or disregarded; advanced, commencing, or possible phthisis are looked upon as identical, and the therapeutic treatment to be adopted is invariably expressed in the clear and concise formula: "Reside in a warm climate, at any rate during the winter, and of the climates which are accessible search for that which is the most warm." Such is the principle laid down, such the ideal sought, and this strange and constant pursuit of heat has such influence that it is found not only in the prescriptions of the physician, but also in works upon clinical medicine of a serious and truly scientific character. Authors, having once begun to compare somewhat analogous places, invariably yield to the above-mentioned tendency, accumulating numbers, arranging stations, and finally deciding upon the superiority of some locality on account of an insignificant difference in the mean heat of the year or season—as if such a difference could be of any importance, when the result depends in reality upon the usual condition of the climate and the relative medical value of the regions. At the same time, unreasonable as the notion is, it results inevitably from the irresistible enthusiasm which directs both the physician and patient in their pursuit of heat. Nothing being said in opposition to this practice, the documents in favour of an indoor life become more and more numerous.

What, then, is the origin of this idea? Simply that a certain fact is assumed to be true. Cold is believed to favour the formation and development of tubercle, it being forgotten that tuberculosis is not bronchitis. On the other hand, it is stated that heat deters the formation and evolution of this product, a fact which is shown to be untrue by the frequent occurrence of phthisis amongst the inhabitants of warm climates. From the above premises it was easy to deduce logically the practical consequences of residence in a warm climate. Unhappily, as regards the foundation of the idea, the premises consist of a supposition which rests upon no proof, and its only basis is a medical error, since the nature of tuberculosis as a diathesis is put out of sight, as well as the certain and in all cases preponderating influence of constitutional malnutrition.

For my part, I have been long estranged from these antiquated notions, and entirely disregard the trite and routine prescription of warm climates. Nor am I more in favour of the opposite theory, that cold climates should be exclusively employed; and, regarding the indication of climate as a separate therapeutic question, I have attempted to find in place of a uniform rule, which is but an illusion and danger, the indications or counter-indications both of hot and of cold climates. I was thus able to establish, in the first edition of my work on Pathology, written in 1870, a fundamental distinction between confirmed and imminent phthisis. Two years later my experience enabled me to extend this idea in my Clinical Lectures to the first stages of the disease, and since that time my attention has been fixed upon this subject, so worthy of interest. After eight years, during which I have unceasingly utilized both the patients' observations and my own personal knowledge of the places in question, I am more persuaded than ever that the fundamental distinctions then made are true, that different indications exist with regard to climate, as in the case of other therapeutic agents, and that the elements of a rational and useful parallelism have been acquired between the pathological and climatic groups. These elements, which form the rules of my practice, will now be explained.

The climatic problem admits of no general and uniform solution drawn from the special effect of heat and cold, or from a supposed connection between tuberculosis and ordinary bronchitis; the solutions, which are various and multiple, depend upon the answers to certain preliminary questions, which should be clearly understood and closely investigated.

First Question.—Has climate of any kind a curative effect upon tubercle? No; the answer is unreservedly negative. It is, therefore, useless to inquire which climates have this effect to the fullest extent, since in no case does such influence exist.

Second Question.—Has every climate an equal effect in phthisis? or do some climates confer upon their inhabitants absolute or relative immunity from the disease? The interest of this question is easily understood. If it is certain that in some special regions tuberculosis affects none of the inhabitants, or but very few when this district is compared with others, it is clear that direct antagonism must exist between these places and the development of the disease. If it is also true that these privileged places only differ from those in the neighbourhood by being less fortunate in the special conditions of their climate, without any difference existing in the way that their inhabitants live, the climatic conditions must be looked upon as causing the immunity mentioned. The question, when thus stated precisely and without ambiguity or equivocation, must be answered in the affirmative. Some parts of Switzerland, Silesia,* and Iceland, some table-lands in the middle zone of the Andes, and the high plains of Mexico, are the most notable of such regions. In many of these the immunity seems to be absolute. In Switzerland, when the height of 4270 feet (1300 *mètres*) is reached, the immunity is almost complete, the official statement of Müller indicating but one case of phthisis in a thousand inhabitants during the five years between 1864 and 1870. In addition to this, if, instead of regarding Switzerland as a whole, certain regions with an altitude of above 4600 feet (1400 *mètres*) are separately considered, freedom from the disease is found to be complete in some, as in the Upper Engadine, the valley of Davos, and the lands which are immediately adjacent.

* See note, p. 287.

Some physicians, who are more ingenious than well informed, have affirmed that the true cause of immunity in these high regions is the great mortality of infants at a tender age, and that this, destroying the weak persons, permits the strong organisms alone to survive, by offering effectual resistance to the onset of tuberculosis. It would be interesting to know the statistical facts upon which this curious theory is founded. The immunity is less complete in the elevated plains of Norway; but at the same time, according to the statement of Lochmann, the disease is more uncommon there than in any other place at the same latitude.

Third Question.—Of the different elements which together constitute the climate, which is the most directly connected with this immunity, and to which is it most immediately due? A decided answer may be given, that it is the altitude of the place. The lower temperature is but an accessory element, in itself wholly insufficient. This is proved by the great frequency of phthisis in the lowly situated regions of Northern Russia, Sweden, and Norway. The difference which exists, as regards immunity from phthisis, between certain places in Silesia * or Switzerland, which are exposed to the same climatic conditions but at a different altitude, establish the same fact. Altitude, then, is the important element; but at the same time it should be recognized that though the effect of the temperature is but secondary, it is not inoperative, as shown by the fact that the lower limit of the protective altitude varies with the latitude. Thus the region of immunity exists at a much higher elevation in the Andes and Himalaya mountains than in Switzerland, a a higher elevation in Switzerland than in Silesia or Styria.†
Thus in Switzerland the elevation is of from 4250 to 4600 feet (*de* 1300 *à* 1400 *mètres*), being of about 1800 feet (550

---

* Silesia is the province in the east of Prussia, and west of Poland.

† Styria is the Austrian duchy to the east of Illyria, and west of Croatia and Hungary.

mètres) in Silesia; and it is this consideration which led to the selection of Brehmer when he chose Görbersdorf * for his establishment, at an elevation of 1830 feet (557 *mètres*). On account of the difference in latitude, the thermal conditions of the climate in the latter place correspond sensibly to those met with in Switzerland at the height of from 4250 to 4600 feet (*de* 1300 *à* 1400 *mètres*). In Styria, judging by the height of the Sanatorium at Aussée,† which is 2300 feet (700 *mètres*), the lower limit of the altitude which confers immunity is higher than in Silesia, but lower than in Switzerland, which is perfectly in accordance with the respective conditions of latitude and longitude. Thus, in the climates which confer immunity, altitude is the most important element, though it must be combined with certain conditions of temperature, since the altitude which preserves in one region will not do so in a latitude which, at the same height, has a warmer temperature.

The study of this question shows decisively the healthful influence of life in the open air, and this fact should not be disregarded. With regard to the agricultural population, the lower limit of the protective altitude is found to descend. Thus Müller's work shows that in these persons the immunity is the same at 2950 feet (900 *mètres*) as in other cases at 4250 feet (1300 *mètres*) or more. On the other hand, an indoor life may notably diminish the good effect of altitude, since in Switzerland, where the industrial districts reach an altitude of 3950 feet (1200 *mètres*), or almost that of immunity, the frequence of phthisis is, according to the statistics of Müller, the same as at the height of from 1650 to 2300 feet (500 *à* 700 *mètres*) in other places. It might be inferred on this account that the protection conferred by altitude is but illusory, and

---

* Görbersdorf is situated in Silesia. See "Health Resorts," J. Burney Yeo, M.D.; "Influence of Climate on Pulmonary Consumption," C. J. Williams; etc.

† Aussée is near Salzburg, in Upper Austria, to the east of Tyrol and Bavaria. See "Handbook for Travellers in South Germany and Austria" (J. Murray), etc.

that the result is really due to the mode of life which the patient leads. This, however, would be a complete error, since it is shown by the same author that in the agricultural population, considered separately, phthisis increases in proportion as the altitude decreases, so as to reach its maximum effect at a zone of from 650 to 1650 feet (200 à 500 *mètres*) in height. It is of undoubted importance how the life is spent, as already observed, but even the most favourable life, namely, that of the agricultural population, is powerless against the injurious effects of low altitude.

Fourth Question.—What climatic conditions are associated in Central Europe with the altitude which confers immunity from the disease? The principal are a temperature which is cold in winter and cool in summer, winds which have a special direction at fixed times during the summer but scarcely exist in winter, and complete pureness of the air. The climate which presents these conditions, including a high altitude, is the type of those which are tonic and stimulant. Hence the conclusion may be formed, that climates at a high altitude having tonic and stimulant effects can alone confer on the inhabitants absolute or relative immunity from pulmonary phthisis.

It would not be reasonable to deduce from these facts that climates at a high altitude have a curative effect upon tuberculosis. It is certain, however, on the other hand, that these climates specially prevent the development of the disease in their inhabitants, and, unless there are special counter-indications, offer to those already affected the most favourable atmosphere which could be desired. The proof of this is furnished by the result of observations continued unceasingly for fifteen years, and also by facts to be now considered, and which are uncontested from being incontestable.

By their tonic and stimulant effect, and by the way in which they promote activity of the nutrition, climates which are highly situated fulfil the indication presented by constitutional debility and malnutrition, which so constantly exist in the complaint. This is already a great result, since no other climate, no other therapeutic means, could attain this object with sufficient effect, in addition to which these climates, on account of their altitude, have a special and direct action upon the condition and function of the lungs. Some difference of opinion still exists as to the whole effect produced upon the organism by the fact of the barometer being always between five and six inches less high than in lowly situated regions. The question whether there is a relative diminution of oxygen in the air breathed, which in some way affects respiration, will long be debated, an idea which, in my opinion, cannot be admitted within the limits of rarefaction which are now being considered. The activity of organic combustion, as indicated by the quantity of carbonic acid contained in the air of expiration, will be no less discussed—a question which, in my opinion, was settled by the observations of Marcet and Chermond, in the sense that mountain air is more beneficial than that of the plains in this respect, that is to say, at the altitudes mentioned above. Besides these facts, however, which are still called in question by some observers, others exist of far more importance in this respect, and which may be looked upon as absolutely certain. My subsequent studies, repeated year by year, have fully confirmed the statements which I made in 1873, in my work on St. Moritz—statements which at that time were based upon observations extending over ten years. Hence I feel myself fully justified in exactly repeating what I then said. Though applying specially to an altitude of from 4900 to 6250 feet (1500 à 1900 *mètres*) in the Swiss Alps, the observation is equally true of altitudes which approach these limits, whether by exceeding them or the reverse; that is, from 6250 to 6500 feet (1900 à 2000 *mètres*), or from 3900 to 4900 feet (*de* 1200 à 1500 *mètres*). The fundamental effects are the same, their intensity alone varying in correspondence with change in the atmospheric pressure.

The diminution of atmospheric pressure at the above-mentioned altitudes usually accelerates to some extent the cardiac pulsations, an effect, however, which in that case is but temporary, and ceases at the end of a few days. On the other hand, during the time of residence the whole circulation is invariably so modified that the blood collects in the superficial parts, the capillaries of the skin being turgescent, while the integument becomes of a reddish violet colour, as do the mucous membranes at the upper part of the body, and notably of the mouth and tongue. After some weeks, this increase of blood in the superficial circulation produces deeper pigmentation of the skin. Since this pigmentation is most marked in regions which are exposed to the sun, it might be supposed due to solar radiation. The same change, however, occurs to a less extent in parts covered by clothing, which plainly indicates the true cause. In some cases which are less frequent than might be supposed, the occurrence of epistaxis shows that some change has occurred in the apportionment of the blood.

The incessant flow of blood to the peripheric parts keeps the viscera in a condition of relative anæmia, which, from being slightly pronounced, produces but favourable results. The cerebro-spinal functions are carried on more actively and with greater facility, the mind is unoccupied and clear, the power of locomotion is increased, the respiration is much facilitated, though the mode of performing it is greatly changed, as will soon be explained. These organic changes produce in the patient a sensation of increased strength as compared with that which he has in his ordinary condition. With a feeling of activity and cheerfulness, there is increased capacity for physical exertion, and the final result of these changes is improvement in the nutrition and restoration of the organic powers.

The rarefaction of the air produces two changes in the

respiratory functions which give rise to important modifications. During rest, the frequency of the breathing may be increased by from three to five inspirations per minute, as in my case, while in all patients the respiration is deeper or more ample. The reason of this is that in a rarefied atmosphere a larger respiratory capacity and more respiratory absorption are required in order to maintain in the lungs the quantity of air necessary for carrying on with increased activity the operations of hæmatosis and nutrition. A slight increase in the number of inspirations, even if constant, could not produce this result, to effect which the capacity of the lung must be increased, certain regions of the organ being then brought into use, which may be termed inactive, and which in the ordinary conditions of life do little towards expanding the lung. These regions are its upper parts. Since, however, the atmospheric pressure is diminished, such increased participation of the lung in the act of inspiration necessarily implies increased action on the part of the muscles which produce expansion of the chest; these subordinate conditions, due in all cases to change of atmospheric pressure, have for their final effect to produce a natural and methodical exercise of the respiratory organs, which, though regular and unremitting, maintains them at the maximum of functional activity without the existence of fatigue.

Thus effects are produced by the active movements of the respiratory organs analogous to those which result from the passive influence of compressed air. In rarefied air the respiratory absorption becomes complete on account of the increased muscular exertion, while in compressed air the increased inspiratory absorption results from the increased pressure, to which the lungs, and the lungs alone, yield passively. This comparison, which, in my opinion, is of interest, suffices to establish the superiority of the first condition as regards the development and regular exercise of the pulmonary functions.

From the energy, constancy, and equality of its effects, from the facility of its application, which occurs as a natural consequence of its use, and from the absence of fatigue, residence in the rarefied atmosphere of the high places indicated is the ideal of aërotherapeutic treatment. The important part which this method takes in the treatment of pulmonary phthisis has been already pointed out, but other consequences of no less value must yet be mentioned.

One of the effects produced by altitude, namely, the diminution of blood in the viscera, and its increase in the superficial parts of the body, has already been mentioned. This relative anæmia, in which the lungs, like other internal organs, participate, adds powerfully to the good influence of respiratory activity by facilitating the pulmonary circulation, removing congestion, should it exist, and preventing any fresh occurrence of hyperæmia. It is thus shown that the prejudice which attributes a tendency to hæmoptysis to the influence of residence at a high altitude is mistaken, since with respect to the heights which have been considered observation has established two facts: firstly, the almost constant absence of hæmoptysis in patients who reside there, and secondly, the cessation of such hæmorrhage in those who suffer from it, even during the days which immediately precede their arrival. In my work already quoted, I have mentioned a person of Modena, affected precisely in this way, and in whom the hæmoptysis ceased from the moment that he reached St. Moritz. Since then I have closely observed the son of a rich manufacturer at Lille, who passed three successive summers in the Engadine without being once affected by hæmoptysis when residing in that region, while during the rest of the year the attacks were frequent.

That these would be the results of observation might have been foreseen, since the opposite hypothesis has no other foundation than an erroneous interpretation of the effects of altitude upon the circulation of the blood. The fact of the blood flowing towards the superficial parts and extremities is acknowledged, but the fault is made of ranging the lungs among the superficial organs, and thus explaining the supposed hyperæmia and tendency to hæmorrhage. Experience, however, has shown that during the special condition of the circulation which is associated with a low barometer, the lungs should be regarded as deep organs, and as sharing their condition of relative anæmia. The researches of Poiseuille and Volkmann have in fact shown that the amount of blood in the thoracic organs is in direct proportion to the atmospheric pressure, the pressure of the air breathed being understood to be the same as that of the air which weighs upon the surface or the body. This is the fundamental difference which exists between residence in a rarefied atmosphere, and respiration of the rarefied air in a pneumatic machine, while the body is exposed to the pressure of the normal atmosphere. The conclusions which have been formed with regard to the occurrence of hæmoptysis are, therefore, without foundation.

Mountainous climates at the height of from 4000 to 6200 feet (*de* 1500 *à* 1900 *mètres*) have in reality a double effect: firstly, a general one, by which the constitution is restored to a healthy condition; secondly, a local one, by which the activity of respiration is increased to a maximum degree, while the lung is protected from the effects of congestion or hyperæmia. Climates which, on account of their more northern latitude, present analogous conditions of temperature at a lower altitude, produce the same tonic effect under these conditions. They have not, however, the same mechanical influence upon the lungs, this being entirely due to the barometric pressure, that is to say, to the altitude of the place. The limits frequently indicated are, in my opinion, necessary, that the altitude may produce its whole effect, but are not so highly situated that any bad consequence is to be feared, while another particularity of great importance connected with the altitude mentioned is the dryness of the air, which is therefore ill suited to development of the lower organisms, its purity being complete.

It would be difficult, as must be allowed, to conceive any climatic influence which is better adapted to effect the object in view. Such a climate is opposed to constitutional malnutrition, improves the local condition, removes the inaction of the lung, and prevents the occurrence of pulmonary congestion, which has the special power of developing lesions or of aggravating those which exist. The fundamental and constant indications so often mentioned are from first to last fulfilled by the climates, and it is surprising that facts which should be universally known did not long ago produce a complete reform in medical practice.

The principle upon which I act is the following: so long as the condition and disease of the patient allow the reverse indications, which will soon be mentioned, to be neglected, the mountainous climates should be utilized, the degree of their power, that is of their altitude, depending upon individual conditions. Consequently these climates are unreservedly adapted to the prophylactic and initial period of ordinary phthisis.

In addition to this, I consider that with regard to individual cases of pulmonary phthisis, and the climates considered as respects its treatment, a new and special division should be made, and the following, in my opinion, should be the one adopted:—

1. Climates which are highly situated with but slight atmospheric pressure.

2. Climates of the plain with moderate, or somewhat less than moderate, atmospheric pressure.

The real types of the first group are stations at the height of from 4900 to 6250 feet (*de* 1500 *à* 1900 *mètres*), but mountain residences should be placed in this group at as low an altitude as 3300 feet (1000 *mètres*) in our latitude, or even 1650 feet

(500 *mètres*) in countries which are farther north. If, in fact, they have not to the same extent either the special effect which results from diminution in the atmospheric pressure on the one hand, or dryness and exceptional purity of the air on the other, there are still certain meteorological conditions which strengthen and fortify the constitution in the same way as mountain air. Such a difference in the intensity of their effects may enable them to fulfil the varying indications presented by the individual patients.

In the second group, which contains climates of the plain with moderate barometric pressure, all places are included with an altitude of less than 1300 feet (400 *mètres*).

The above division is evidently the most important and rational, as can be shown without difficulty. Even if this, however, is not the case, and the climates with moderate barometric pressure resulting from their temperature, the quality of the air, or other conditions, have the same favourable influence upon the general condition, nutrition, and strength of the organism, the patients would still be deprived of the healthful action of a rarefied atmosphere. On that account alone, even if all else were the same, such climates differ from those previously mentioned, and should be clearly distinguished from them. This indispensable condition is fulfilled by the above division, founded upon the principal differential points of the various climates as regarded from the special point of view now under consideration. In reality, however, the difference between the two groups is not confined to the special effects of their altitude, but is also seen in their general effects upon the organism, a second distinctive character which also justifies this division. In fact, the climates with moderate atmospheric pressure, of which the second group is composed, comprising the low-lying watering-places of Switzerland, the Tyrol* and Austria, on the shores of the Mediterranean Sea,

---

* The Tyrol is in Austria, to the south of Bavaria, and north of Lombardy.

in the south of France, in Italy, Greece, Spain, and Portugal, to which those of Madeira,* the Canary Islands,† Morocco,‡ Algeria,§ and Egypt ‖ should be added, do not exercise the same tonic effect upon the organism as the climates at a high altitude—this, in fact, being often absent. On the other hand, their action is often quite the reverse, and tends to produce debility and depression. Thus the above division is most important and fundamental, since the pronounced characters of the climates are contained in one statement as regards their action upon the organism, and therefore their therapeutic effects.

It results from this comparative review that the climates with moderate pressure, whether mountainous or maritime, fulfil none of the ætiological indications drawn from the nature of the disease. The mechanical action which results from a rarefied atmosphere is absent, the tonic and fortifying effect is insufficient, and the special purity of the atmosphere peculiar

---

* See note, p. 243.

† The Canary Islands are in the North Atlantic Ocean, off the west coast of North Africa. See Teneriffe, p. 377.

‡ Morocco, the empire in Africa on the shore of the Mediterranean Sea, and of which the capital town bears the same name, is considered by some a beneficial winter residence   See "Health Resorts," J. Burney Yeo, M.D.; "Handbook to the Mediterranean" (J. Murray); "Morocco: its People and Places," Edmondo de Amicis; etc.

§ Algiers is the capital of Algeria, on the north coast of Africa. The principal suburb of the town is termed Mustafa Supérieur. See "Dictionary of Watering-places," part ii. (L. Upcott Gill); "Curative Effects of Baths and Waters," Julius Braun; "Winters Abroad," R. H. Otter; "Health Resorts," J. Burney Yeo, M.D.; "Handbook to the Mediterranean" (J. Murray); "Handbook for Travellers in Algeria and Tunis" (J. Murray); "Health Resorts for Tropical Invalids," Moore; "Principal Southern and Swiss Health Resorts," W. Marcet, M.D.; "Influence of Climate on Pulmonary Consumption," C. J. Williams; "European Guide-book for English-speaking Travellers," Appleton; etc.

‖ Egypt, on the north coast of Africa, and its metropolis Cairo upon the eastern side of the river Nile, are used as winter residences. See "Curative Effects of Baths and Waters," Julius Braun; "Principal Southern and Swiss Health Resorts," W. Marcet, M.D.; "Health Resorts," J. Burney Yeo, M.D.; "Winters Abroad," R. H. Otter; "Handbook to the Mediterranean" (J. Murray); "Handbook for Travellers in Egypt" (J. Murray); "Egypt and the Nile considered as a Health Resort," John Patterson; etc.

to high altitude does not exist. In consequence these climates differ from the others, in fulfilling indications of but secondary value, and which are not of great importance either in themselves or on account of the period at which they appear.

These indications of secondary value will now be considered.

To appreciate them completely, and without underrating or exaggerating their value, the possible effect of such climates upon tuberculosis and phthisis should be well understood. This may be shown to be wholly indirect. Like other remedies, they have no curative effect upon the tubercle, nor do they prevent the development of fresh tubercular products. At the same time, having a fresh temperature in the winter, which is either temperate or warm, with more or less equality of temperature, they may produce a favourable effect upon pre-existing bronchitis or pulmonary catarrh. The bronchial tubes are to some extent protected from inflammation, while confinement to the room being unnecessary, several hours can be spent daily in the open air without danger of bronchitis or pneumonia, which would be certain to occur at this period of the disease should any attempt be made to live in a more rigorous or variable climate.

Such, then, is the whole effect of stations with moderate atmospheric pressure. While producing some benefit, their action should not be exaggerated. The essential advantages yielded by the climates themselves without the employment of medicinal remedies are that they enable life to be spent out of doors, and give the healthful effect of the sun and open air upon the organism, without the risk of chills or other accidental complication.

This being the case, is there any time during the progress of the disease when this indication becomes, as far as residence is concerned, of preponderating importance? Unfortunately such a period always arrives, and except in those rare but undoubted cases in which the disease is arrested either at its very onset or after so short a time that the question of residence at a high altitude need not be taken into consideration, this moment will certainly occur in the case of every patient. Thus it will be seen that, notwithstanding the relative inferiority which should be assigned to climates with moderate atmospheric pressure, their importance is by no means inconsiderable in the treatment of phthisis. At the same time, such treatment should not, in my opinion, be exclusive, and should hold but a secondary place both in importance and with regard to the period at which it is put into practice.

This fact being recognized, another question arises of considerable practical importance. At what time in the disease should this point be settled, or, in other terms, at what period does the special benefit conferred by climate become of preponderating importance although fulfilling but a secondary indication? This question will be answered by means of conclusions drawn from my own personal observations.

It is important to recognize as far as possible that a difference exists in the action of different climates. Thus, when a climate with moderate atmospheric pressure is substituted for one which is more highly situated, the air which is breathed will be less pure, the mechanical effect of the altitude will be lost, as well as the invaluable action which highly situated residences exercise upon the constitution and nutrition of the organism. The latter fact should on no account be forgotten, since even if in the large group of stations at a low altitude those alone are considered which have a tonic effect, their influence in this respect will still be far removed from the truly effective action of a high climate. Such a change is, therefore, to be always deeply regretted, and should be deferred as long as this can be done without injury to the patient, the benefit of favourable conditions which cannot be otherwise supplied being thus retained for as long a time as possible.

These general rules having been now mentioned, the different indications which show that this radical change should be made in the climatic treatment will now be considered. Dogmatic statements are impossible for the reasons mentioned with regard to mineral waters. With respect to the application of this form of treatment, the indications have a purely individual character; the decisions made, notwithstanding the apparently fixed character of the elements upon which they are founded, varying according to the patient, the general precepts which experience has shown me to be applicable in all cases without reserve will be alone stated. Rules may perhaps be deduced therefrom, but these would vary according to the different cases, and it would be impossible to consider them absolute.

Certain pathological phenomena will first be considered whose signification is not in all cases the same.

Catarrh of the apices occurring at the onset of the disease, and which accompanies or often precedes the first formation of tubercles, does not counter-indicate residence in a high climate; far from this, it specially requires such treatment so long as the other circumstances of the patient give the same indication.

In the later catarrh which denotes softening of the tubercles, that is to say, passage from the first to the second period of the complaint according to Laennec's classical division, there is no reason for abandoning this mode of treatment. Even when the height, either at first or ultimately, exceeds the known limits of the so-called tuberculous zone, even when, as the height of the residence is increased, the catarrh acquires a more permanent character, the climate of the plain is not necessarily indicated. This, in fact, is only the case when certain circumstances exist, which will be the more readily mentioned since their importance has until the present time been completely misunderstood. The real and only counter-indication in any given case, and which often occurs in practice, is the following :—

When the patient has already lived for some time in a highly elevated climate no inconvenience, but, on the other hand, advantage, occurs in his remaining there, so long as none of the counter-indications exist, which will now be taken into consideration. If, however, the patient does not live at a high altitude, and an opinion has to be formed at the beginning or during the course of the winter, the condition of things is entirely different. Not only will the patient be unaccustomed to the fresh climate, but will have to live there under conditions which are specially disadvantageous to the inhabitant of a valley. It is to be feared that one of the first effects of this unseasonable change would be the occurrence of bronchitis, and the increase of pre-existing catarrh. Thus it should be ascertained that the patient is accustomed to such a climate, or if not, would be so when the change is first made.

When the contingency occurs which has just been mentioned, namely, the existence of secondary catarrh, associated with softening tubercle in a patient unaccustomed to high climates, and the question of residence at a high altitude has to be settled, such treatment should not, in my opinion, be employed. The most urgent, the most real indication, so to speak, is to prevent as far as possible any complications which may affect the bronchial tubes, and on this account, since the patient is unaccustomed to such a climate, it would not, in my opinion, be right that he should incur the danger of sudden exposure to a low temperature. In these cases the physician is not free to act in what would otherwise seem the most beneficial manner, and though climates at a high altitude are usually to be recommended, an exception should be made in patients of this class. Whatever may be done at a later period, in my opinion it is better at the present time to employ climates in this group which are placed at a lower altitude, or those with a moderate degree of atmospheric pressure. Thus in such conditions, when catarrh, which is due to softening of the tubercle, becomes permanently extended beyond the limit of the tuberculous zone, a positive though transient indication exists that climates with a moderate degree of atmospheric pressure would be most beneficial.

In the same circumstances, that is to say when the patient is unaccustomed to high climates, the frequency, and still more the increasing frequency, of acute complications is a second indication of the same fact. The individual varieties in this respect are most remarkable. Some patients pass through the first and greater part of the second stage of phthisis without any acute complication, and in such cases highly situated climates may be long utilized. Others, from the first moment that phthisis is known to exist, suffer from bronchitis, or pulmonary congestion, and when the period of softening occurs these complications become more frequent, in many cases without obvious cause. It is very possible that during the interval between the attacks no increase occurs in the fundamental lesion; this, however, is rarely the case, and even when it is so, the tendency to acute complications, which becomes more and more pronounced, would counter-indicate the sudden adoption of a high climate as a winter residence.

The facts just considered have by no means the same signification when the question has to be decided in summer, or at the beginning of autumn. In such a case they merely impose the necessity of a very gradual transition from the valley to an elevated place which is to be the patient's definite residence, and do not counter-indicate the use of high climates, which on the contrary would, in my opinion, be of special benefit. The patient becomes accustomed to the new atmosphere under such conditions. Residing in a high climate without causing any danger to the patient has now a favourable effect upon the catarrhal complications and tendency to congestion, and when the winter comes, having been inured to the atmosphere during the preceding season, all the good effects of the climate, both general and local, may be obtained without more liability to the onset of bronchitis than in the valley.

The distinction which has just been stated is new, strange as this must appear, owing to its fundamental character. If this be remembered, many of the apparent counter-indications to the use of high climates will be removed. These should not be classed with the so-called winter stations, since, though the winter may be spent at them, it is not in that season that the climatic treatment should be put into practice. When once the existence of the disease is certain, and it is not the preventive treatment alone which has to be considered, it is in the summer or autumn that the patient should become accustomed to the place, and by taking this precaution, elementary as it seems, he will derive the most benefit and least disadvantage from residing in such a climate. This practical rule has also the great advantage of reducing to their true value the counter-indications to these climates which will now be considered.

During the prophylactic period no counter-indication exists but those pathological conditions which are unconnected with phthisis, and which in any case would exclude residence at a high altitude, and specially when this exceeds 3950 feet (1200 mètres). Such are general emphysema, and disease of the heart or blood-vessels, etc., as has been already mentioned with regard to St. Moritz.* In every case, however, and specially when signs of nervo-vascular excitability exist in a patient, the change to a highly situated residence should be made gradually and with care, the importance of this having been already stated.

When tuberculosis is known to exist in its ordinary and most important form, the chief counter-indication to residence

* See note, p. 250.

in a high climate is furnished by the mode in which reaction occurs. Whatever the stage of the affection, the course of the symptoms, and the lesions may be, should the patient be individually of an excitable nature, high climates must on no account be taken into consideration. Certain to be injurious, they may produce a harmful effect, either by leading to fresh complications, or by aggravating those of a distressing nature which already exist, especially nervous excitability, fever, and insomnia. Consequently, and without considering the other elements of judgment, these climates are specially beneficial to patients with a torpid reaction, though not to them alone. Between the two opposed types of reaction an intermediate group exists, to which the greater number of patients undoubtedly belongs, and of which the characters, being less pronounced, have a more doubtful meaning. The reaction of patients in this group may be termed indifferent, being rather of the torpid character in some patients, while in others it is of a more active type. No definite rule can be given in this ill-defined group, and the decision in each case must depend upon the individual character of the person. The history of the patient before the onset of tuberculosis must be carefully learnt; his mode of life and ordinary residence, his education and character, must be taken into consideration. The effects which the affection produced at its invasion, and the course which it took, must be no less fully ascertained, and thus the question should be solved whether high climates would prove beneficial or injurious to the individual patient. This is a most difficult therapeutic problem. It is not difficult to assert that if a patient is found to be of the torpid type the watering-places situated at a high altitude and with slight atmospheric pressure would be most appropriate, whilst if the reverse is the case, a valley with increased pressure should be chosen. This, however, is but a statement of which the practical application must be the result of experience.

Another counter-indication, quite as absolute and easily understood, is what may be called the wasting period of the disease (*la phase consomptive de la maladie*). Whether occurring early or late in the complaint, it should, in my opinion, unreservedly exclude climates at a high altitude To keep, however, within its true limits, the bearing of that observation, the consumptive stage, must not be confused with that of fever; the question should be solved by considering not only the symptom fever, but also the deterioration of the constitution as a whole. The distinctions already established with regard to pyrexia are now of great value. The fever, which is combined with acute complications, being due to inflammation of the bronchial tubes or lung, and which begins and ends when they do, supplies no precise indication with regard to climatic treatment, except, perhaps, the elementary and general one, that all change of residence should be deferred until the complication has ceased to exist. The choice between the two groups of climates will then be regulated by the conditions which determine it in every case.

What I have just said about inflammatory fever is also true of that due to the softening of tubercle, or formation of caverns. Whilst such a complication is in the acute stage, any change of residence would, in my opinion, be a mistake; but when this ceases the condition of the patient will be unaffected so far as the choice of a dwelling is concerned, the occurrence of a febrile complication not having imposed any definite line of treatment. The same may be said of fever due to the formation of tubercle combined with secondary granular formations. On the other hand, the so-called hectic fever, being unaccompanied by any notable change in the local lesions, should be ascribed to the absorption of noxious products, and is therefore of a different nature, as is the wasting which is so often associated therewith. Of whatever type it may be, fever is an element in the consumptive phase of the disease, of which it is plainly characteristic, and though the emaciation may be as yet but little pronounced, counter-indicates the employment of high climates. At the same time, that no hurried decision may be made, it is my habit in such a case to try first the effect of antipyretic agents; if, under the influence of such treatment, the fever subsides, the counter-indication is less urgent, and a high climate may be utilized without danger to the patient, so long as he becomes accustomed to it in summer, the necessity of which has been already shown. When, on the other hand, the fever is not controlled by antipyretic agents, all doubt ceases as to the employment of high climates, which should be completely and definitely abandoned.

The preceding counter-indications, drawn from the mode of reaction and general condition of the patient, are the most notable, as may be again observed. At the same time, certain particulars connected with the local lesions of the disease are at least of equal importance, and these experience enables me to describe. In each stage and in every form of the disease, ulceration of the larynx in tuberculosis should completely and unreservedly exclude the use of climates with slight atmospheric pressure, and usually those also which are severe and unseasonable. Ulceration of the intestine should have the same effect, while on the other hand diarrhœa, which is simply due to intestinal catarrh, or gastro-intestinal dyspepsia, is favourably and rapidly modified by the effect of elevated climates, and specially by the extreme places of this group.

The extent of pulmonary lesion should be taken into serious consideration. If limited to the apices, the indication furnished by it may be neglected. When, however, a large part of the lung is affected, when the lesions are bilateral and extend to a considerable depth so as to suppress a large portion of the surface in which hæmatosis is effected, they are of the highest importance. Residence at an extreme altitude becomes impossible, since hæmatosis in such a rarefied atmosphere would be insufficient to meet the requirements of the organism, and the patient would undoubtedly be subject to persistent dyspnæa. If the other circumstances and particulars of the case positively indicate treatment by residence in a high climate, the places at the lowest altitude in the group should be alone inhabited for many months, and then only by way of trial. This course should certainly be adopted, since if the trial succeeds the lesions will perhaps become less extensive during this time, so that eventually the patient may be able to reside in climates of higher altitude.

From this point of view the extent of the lesions is of far more importance than their character. Thus the existence of caverns alone is not in itself a counter-indication to residence at a high altitude, everything depending upon their number and size; that is to say, upon the extent to which the surface is diminished in which hæmatosis takes place. This should be looked upon as a foundation upon which the decision is to rest, and it must be acknowledged that some cavities are favourably modified by the effect of a high climate. Such are those of small size, in a stationary condition, containing liquid which produces no irritation, and therefore unassociated with any inflammatory complication in the neighbouring parts. In these conditions expansion of the healthy portions of the lungs, under the influence of a rarefied atmosphere, may concur effectively in the occlusion and cicatrization of the existing loss of substance. Upon this point my conclusions are quite in conformity with those of Powell.

Amongst the local lesions of ordinary phthisis, the pneumonic foci deserve special attention; that is to say, not those of intercurrent pneumonia which end in complete resolution, but those which remain when the pneumonia itself has ceased to exist. This disease, whenever it is followed by acute symptoms, which recur at short intervals, counter-indicates residence in a high climate. When, on the other hand, there is a complete calm after the passing storm, the symptoms of disease not recurring at all, or only at long intervals; when, in fact, nothing remains but condensation of tissue due to the inflammatory process without any sign of acute disease, this counter-indication cannot be said to exist. It has been observed, in fact, that the rarefied atmosphere which exists at the highest points in the first group of climates induces and favours the absorption of any accidental infiltration. When in such a case the condition of the patient permits this resource to be utilized, it is, in my opinion, of great advantage to the patient.

The hæmoptysis which occurs in the first or second stages of phthisis is favourably affected by residence at a high altitude, as was clearly shown when the changes in the circulation produced by notable rarefaction of the atmosphere were being considered. From this mechanical fact, however, no general law can be deduced. As in the case of other patients suffering from phthisis, those in whom hæmoptysis occurs present the two different forms of reaction which are termed active and torpid. In the first, though the other physical symptoms may seem to indicate residence at a high altitude, this would be, in my opinion, prejudicial. Experience has proved the truth of the fact so conclusively that it cannot be doubted, while in the reverse case high climates are undoubtedly the most beneficial which can be recommended.

Of cases which might be adduced in support of propositions which are so absolute, two will be selected which have been observed by me within the last few years. In one, the patient was the son of a manufacturer at Lille, and has been already mentioned; in the other, a young man from Russia, of the same age. In both the tuberculosis took the same form, presented the same anatomical changes, and both cases were characterized by frequent attacks of hæmoptysis attended by pyrexia. The two cases, in fact, were analogous both as regards the local lesions and the attacks of hæmoptysis, while the pathological changes seemed to be identical. Notwithstanding this, the young man of Lille continued to be affected by hæmoptysis in every climate except at the high altitude of St. Moritz, while the Russian, who of his own accord tried the effect of numerous watering-places situated at different altitudes, obtained relief from his hæmoptysis at Madeira* alone. At the same time, there was no difference between the two patients except in the mode of reaction; the Frenchman, in whom it was of the torpid type, being relieved of his hæmoptysis at a high altitude, while the Russian, in whom excitability was the marked character, was only cured in the climate of Madeira. The latter, after being free for some months from this complication, was so imprudent as to leave the island and visit Pau,† notwithstanding the unquestionably sedative qualities of that station, and was then again affected by the same complication.

This fact, of considerable practical importance, should therefore be remembered: notwithstanding the precise character of the atmospheric effect upon the pulmonary circulation, hæmoptysis does not itself furnish any uniform indication with respect to climatic treatment. The coexistence or absence of pyrexia affords but vague indications, and the decision as to climate should entirely depend upon the power of reaction on the part of the patient. This is undoubtedly a

* See note, p. 243.
† Pau is a French town, in the department of the Basses Pyrénées. See "Dictionary of Watering-Places" (L. Upcott Gill); "Curative Effects of Baths and Waters," Julius Braun; "Health Resorts," J. Burney Yeo, M.D.; "Principal Southern and Swiss Health Resorts," Wm. Marcet, M.D.; "Wintering in the Riviera," W. Miller; "Handbook for Travellers in France" (J. Murray); "Dictionnaire Encyclopédique des Sciences Médicales," Paris; "The Curative Influence of the Climate of Pau," A. Taylor, M.D.; "The Climate of the South of France," C. J. Williams; "Influence of Climate on Phthisis," C. J. Williams; "European Guide-book," Appleton; "Guide to the South of France," C. B. Black; etc.

most delicate medical question, in which an error in either direction might have the most serious consequences.

In the pneumonic form of phthisis (the special form of the disease to which this term is applied will be remembered) climates at a high altitude are absolutely counter-indicated. The infringement of this rule would probably have as a consequence the return of acute symptoms.

It need scarcely be added that when the disease is known to exist, the necessity of increasing the height of the residence by gradual changes when an altitude of 3950 feet (1200 *mètres*) is reached is even more imperious. This is necessary, both from the certain effect of the climate upon the nervous and cardiac excitability, and from the disturbance in the circulation produced by rarefaction of the air; and though the new climate may be beneficial to the patient, it is dangerous, and undoubtedly so, to impose it suddenly upon the organism. In considering the waters of St. Moritz, the complications were mentioned which might result from neglecting this precaution, and need not be here reconsidered.

Though the counter-indications to residence in a high climate are somewhat absolute, they should be neglected in the case of patients who habitually reside there, and have become consumptive from living in the plain. In these circumstances, a return to the high climate is absolutely necessary, and outweighs all other considerations. Nothing can replace its marvellous effects, as I have myself seen in the case of patients residing in the Upper Engadine.

The counter-indications furnished by the disease itself should not, of course, be the only ones taken into consideration when the question of residence in a high climate has to be settled. The conditions which in every case are ill adapted to such a course must also have their weight. These, as has already been indicated, are generalized emphysema, and disease of the heart or blood-vessels.

The general indications and counter-indications of the two fundamental groups into which the climatic stations are divided, namely, into high climates with slight, and climates of the valley with moderate atmospheric pressure, have now been discussed. As to the selection made in each of these groups, specially in the second and most numerous one, this must depend upon the individual peculiarities of each patient as well as upon the form of the disease, each of which will be subsequently discussed.

Such being the new point of view from which my experience has led me to regard the question of climatic treatment, it will be understood how little importance I attach to an increase or reduction of some tenths of a degree in temperature, or to the summer occurring at an earlier or later time of year, in the place considered. This will be still better understood when the relative importance of the different meteorological changes has been discussed in connection with the watering-places of the temperate zone; it will then be seen that the difference in temperature, of which so much has been said, is but of secondary consequence. Founding my opinion upon a direct knowledge of the places, as well as upon what has been observed in my patients, and regarding the climatic stations from a general point of view, I have been specially struck by the radical difference already mentioned, which has led me to divide them into two groups dependent upon the altitude of the places, a distinction which implies a no less absolute difference in their therapeutic effect.

The truth, in fact, is, that high climates with slight atmospheric pressure have not only a direct and restorative effect upon the constitutional condition, but also exercise a no less healthful influence over the functions and circulation of the diseased organs. Consequently, climates bear an undoubted part in the question of treatment, and, their effect being active and positive, are real therapeutic agents.

Mild climates with moderate atmospheric pressure exercise no direct action upon the functions and circulation of the respiratory organs. They have little or no restorative effect upon the nutrition and strength of the body. They act indirectly by preventing accidental complications, maintaining the *statu quo*, and enabling these benefits to be combined with life in the open air. Consequently, since they take no active part in the treatment, the climates should be looked upon as passive or preservative, being rather witnesses of than agents in the therapeutic treatment.

Their effect, however, is by no means unimportant, since while such preservation continues, as it does for some length of time, beneficial changes may occur, and the patient be thus enabled to derive benefit from all the favourable contingencies which the natural course of the disease allows to take place. At the same time, this indirect and passive influence is undoubtedly inferior to the direct and active effect produced by high climates, and the great difference which exists between the physiological action of the two groups is also found in their respective therapeutic effects.

It is evident, therefore, that, however climates are regarded, the above distinction is the truly fundamental one, and the necessary guide in medical practice. Nor is it less evident that the passage from the first to the second group of climates should be made as gradually as possible, since a treatment which is powerful and direct is then being abandoned for an influence which is neither so energetic nor so immediate. It is only to be regretted that the personal conditions of the patient, or perhaps the course and modifications of the disease, often give rise to phenomena which can alone decide the selection, removing any freedom of choice, and obliging climatic treatment which is active and effectual to be replaced by such as merely prevents extension of the disease.

CHAPTER XII.

CLIMATIC TREATMENT (*Continued*).

General direction of the climatic treatment—Ordinary plan of changing residence—New plan of fixed abode—Its causes and advantages—Mode of application.

Conditions which should be filled by a medical station intended for phthisical patients — Sanitary conditions—Hygienic conditions — Residence of the patient—Therapeutic resources.

Application of the climatic treatment—Climates of the first group—Davos and the Engadine—Analogy of and difference between the other stations of the same group—Preponderating importance of altitude.

Indications and counter-indications of highly situated places—Necessary precautions—Difficulties in the therapeutic selection.

My classification of the climates in connection with the treatment of pulmonary phthisis has been explained and discussed. It has been shown that from a therapeutic point of view the fundamental indication should be drawn from the atmospheric pressure rather than from the temperature, and that the climates with slight and those with a moderate degree of atmospheric pressure should be specially separated. The climates belonging to the first group are all severe in winter, but differ notably from each other in atmospheric pressure, which diminishes, as the height of the places varies from 1650 feet (500 *mètres*), as in the more northern latitudes, to more than 4900 feet (1500 *mètres*). The stations at the extreme ends of this group are typical both as regards their meteorological conditions and curative effects. The altitude of the climates of the second group in the latitude of our country is comprised between 1300 feet (400 *mètres*) and the level of the sea. These climates bear the following points of resemblance to each other; the mean atmospheric pressure is but moderate, while the temperature is cool, temperate or warm in winter, hot or scorching in summer.

The respective action of these two groups of climates was also discussed, and the conclusion formed that climates with slight atmospheric pressure, and notably those at the extreme end of the group, are specially efficient and curative, while those with greater pressure are solely and passively preventive. This being explained, the general indications of the two groups were considered, and the different conditions mentioned which counter-indicate the employment of high climates.

In what way should the fundamental conclusions arising from these considerations be applied? what practical rules do they furnish? These questions will now be discussed, though before doing so two other important questions will be considered, which, unfortunately, are but too often neglected. The first regards the general management of the climatic treatment in patients affected by phthisis; the second refers to the different conditions that a watering-place should fill, to whichever group it belongs.

With regard to the first of these questions, the general management of climatic treatment, the method which I recommend differs from that which is usually adopted, and which may be summed up in a few words, namely, variation of the residence according to the season. Such is the plan followed, one place being occupied in summer, another in winter, often a third in the spring, and a fourth in the autumn. This is the method, if it is right to designate by such a term a plan of action which can only be inspired by indifference, routine, or scepticism.

No error, as I believe, could be more injurious to the patients, nor can I find any cause, or even pretext, to justify the successive changes which are imposed upon them. Urged by the fatalism which, in the form of a prescription, regulates his movements, the unhappy patient wanders without reprieve from one station to another; scarcely has he begun to experience the beneficial effects of rest in an appropriate residence, when the requirements of constant change in correspondence with the season come again into force, and pitilessly drive him therefrom. His departure is necessary, and the difficult and often dangerous work of acclimatization must again begin. Thus he moves, and is always moving from country to country, the wandering victim of phthisis and medical tyranny, until the wave which has borne him on casts him adrift in a last effort, and leaves him miserably stranded like an abandoned waif in some chance hostelry. Fortunate will he indeed be if not compelled to make a fresh journey during the last days of his life, a course which it will be attempted to justify by reasoning of the most plausible appearance, but of which the real cause is a culpable desire to prevent any addition to the necrological roll of the station.

Applied in this way, climatic treatment is but a formality, a succession of journeys supposed to be more or less appropriate, but in reality harmful; while the chance of any therapeutic advantage is totally lost.

This form of treatment, however, is truly powerful and effective when based upon other principles than that of heat in winter and cold in summer, and applied in quite a different manner. Such has been my theory for a considerable number of years, and daily experience has confirmed the truth of these views.

In my opinion, climatic treatment is entirely a question of individual adaptation, and the problem is to determine, by means of information acquired from the patient, written documents, and above all personal acquaintance with the places, which residence would be most appropriate to the patient in question. When a practical trial has shown that this suitability really exists, the place should be inhabited for as long a time as possible, with the view of obtaining all the benefit that such a residence can produce. This rule which I have laid down implies a complete revolution in climatic therapeutics, and yet what can be more natural, more simple, I would almost say, more elementary? In such cases the climate is the remedy; if beneficial, why should it be changed? To do this would be most unreasonable. No physician would, as I believe, act so inconsistently in the case of other therapeutic agents; why, then, should it be done with respect to climate? The reason of this mistake is unintelligible unless it be due to the popular division of stations into those of winter and summer. The principle which I adopt and put into practice is of a different nature, namely, to discover a place appropriate to the disease, and then to keep the patient there so long as observation shows that good effects are being produced, and no counter-indication arises. Experience, already of long duration, has led me to believe in the advantage of this system, and that, if it is not always effective, it is at least beneficial, and always preferable to the uncertain and illogical system of change from place to place as the season varies.

Such a method, which I shall call that of fixed, as opposed to the ordinary practice of varying residence, is not only advantageous but absolutely necessary in the treatment of patients by means of high climates.

In the adoption of this plan, there is, in fact, no idea of a summer or winter residence, but that of a rarefied atmosphere, and of the climatic conditions associated therewith. The question, in fact, refers to a remedy, whose action is slow but certain in appropriate cases, and the application of which should on no account be interrupted so long as it is found to produce beneficial effects. No serious inconvenience seems to be associated with this plan so long as the change from one height to another is effected in the careful and gradual manner often affirmed by me to be necessary. The patient may then be allowed to pass one or two months of each year, either during spring or summer, at the home of his family, always supposing that that home is not situated in a country which is very far south. This, however, is the only concession which, in my opinion, should be made in this special form of treatment, and I always prefer that even these temporary interruptions should not occur, and recommend in consequence that any necessary visit should be made in preference by the family to the patient.

It is not impossible to combine the application of this principle with some variation in the place of abode. If, without any alteration in the altitude of the residence, a place can be found of which the climatic conditions are precisely the same from all points of view, the patient, should such be his wish, might take advantage of this opportunity. Such a change would not only produce in that case no inconvenience, but, by interrupting the monotony of the residence for two or three months, would also carry with it the decided benefit of such healthful distraction and moral satisfaction as is in all cases desirable. Unfortunately, this possibility is not realized by every station in the first group. Davos and the Engadine* alone enable this change to be made in conformity with the above-mentioned conditions, that is to say, without unreasonable change in the atmospheric pressure. Thus, if Davos be taken as an instance, situated at the height of 5108 feet (1556 *mètres*) above the level of the sea, it is obvious that, without ever quitting an altitude of between 3940 and 6050 feet (*entre* 1200 *et* 1850 *mètres*), the

* See note, p. 243.

patient might visit Wiesen,* Parpan,† Muhlen,‡ and Churwalden,§ or, with even more benefit, one of the stations in the Upper Engadine, namely, St. Moritz,‖ Samaden,¶ Pontresina,** Campfer,†† Silva Plana,‡‡ or Sils Maria.§§

* Wiesen is a Swiss village in the Engadine. See "Handbook for Travellers in Switzerland" (J. Murray); "Handbook for Travellers in Switzerland," K. Baedeker; etc.

† Parpan is a Swiss village between Coire and the Engadine. See "Handbook for Travellers in Switzerland" (J. Murray); "Handbook for Travellers in Switzerland," K. Baedeker; "Hachette's Diamond Guide to Switzerland;" etc.

‡ Muhlen is a Swiss village, situated between Coire and St. Moritz. See "Handbook for Travellers in Switzerland" (J. Murray); "Handbook for Travellers in Switzerland," K. Baedeker; etc.

§ Churwalden is a Swiss village in the canton of Grisons. See "Health Resorts," J. Burney Yeo, M.D.;" "Curative Effects of Baths and Waters," Julius Braun; "Handbook for Travellers in Switzerland" (J. Murray); "Handbook for Travellers in Switzerland," K. Baedeker; "Hachette's Diamond Guide to Switzerland;" "The Alps and how to see them," J. E. Muddock; "European Guide-book," Appleton; etc.

‖ See note, p. 250.

¶ Samaden is the principal village in the Upper Engadine, Switzerland. See "Curative Effects of Baths and Waters," Julius Braun; "Health Resorts," J. Burney Yeo, M.D.; "Handbook for Travellers in Switzerland," etc. (J. Murray); "Handbook for Travellers in Switzerland," K. Baedeker; "Hachette's Diamond Guide to Switzerland," "The Alps and how to see them," J. E. Muddock; "European Guide-book," Appleton; "Guide to East Switzerland," John Ball, F.R.S.; etc.

** Pontresina is a Swiss village in the Engadine. See "Dictionary of Watering-Places," part ii. p. 77 (L. Upcott Gill); "Curative Effects of Baths and Waters," Julius Braun; "Health Resorts," J. Burney Yeo, M.D.; "Handbook for Travellers in Switzerland" (J. Murray); "Handbook for Travellers in Switzerland," K. Baedeker; "Hachette's Diamond Guide to Switzerland;" "The Alps and how to see them," J. E. Muddock; "European Guide-book," Appleton; "Guide to East Switzerland," John Ball, F.R.S.; etc.

†† Campfer is a Swiss village in the canton of Tessin. See "Curative Effects of Baths and Waters," Julius Braun; "Health Resorts," J. Burney Yeo, M.D; "Handbook for Travellers in Switzerland" (J. Murray); "Handbook for Travellers in Switzerland," K. Baedeker; "Hachette's Diamond Guide to Switzerland;" "European Guide-book," Appleton; etc.

‡‡ Silva Plana is a Swiss village in the Engadine. See "Curative Effects of Baths and Waters," Julius Braun. "Health Resorts," J. Burney Yeo, M.D.; "Handbook for Travellers in Switzerland" (J. Murray); "Guide to East Switzerland," John Ball, F.R.S.; etc.

§§ Sils Maria is a Swiss village in the Engadine. See "Health Resorts," J. Burney Yeo, M.D.; "Handbook for Travellers," etc. (J. Murray); "Handbook for Travellers in Switzerland," K. Baedeker; "Hachette's Diamond Guide to Switzerland;" "The Alps and how to see them," J. E. Muddock; "European Guide-book," Appleton; etc.

The hygienic and other conveniences, which are so exceptionally good at Davos, exist also in the hotels of the Engadine; and the climatic treatment need not be interrupted for a moment on account of the remarkable analogy between the atmospheric conditions existing during summer in all these places, which in fact belong to the same region as Davos. This fact, therefore, allows a change of residence during summer to be combined with not only apparent but real continuity of the treatment to which the patients have been subjected.

With the exception of these cases, no interruption should, in my opinion, take place in the treatment by climate at the height of 4920 feet (1500 *mètres*) or more.

With regard to prophylactic treatment, this should be continued until the weakened constitution is completely repaired. One year will usually effect this when such treatment is only required on account of the individual conditions of the patient. Eighteen months or two years may be necessary when such treatment is indicated by hereditary predisposition.

No definite limit can be put to the duration of treatment by a climate of high altitude when the malady actually exists, and the character of the patient and form of the disease indicate its use. It should be rigidly continued, with the exception of the temporary interruptions already mentioned, so long as the patient derives any benefit therefrom, and while none of the counter-indications exist which have been already mentioned in making a general review of the subject. The treatment, being appropriate and beneficial, should be unchanged. Nothing can be more evident.

The application of this method to climates of the plain will now be considered.

It might be supposed that this mode of treatment would in such a case be impossible, since the stations of the group, which without exception belong to the class of so-called winter stations, present in the spring, or more still in the summer, conditions of temperature which are directly opposed to continuation of residence thereat. The difficulties, however, are not in any way excessive. With regard to some of these stations, the region to which they belong contains within a short distance places which are so analogous that one can without danger be substituted for the other. The intolerable heat of summer can be thus avoided without any alteration of the climatic influence. Madeira,* of all the stations in our hemisphere, most entirely fulfils this fortunate condition. The places by the river-side at the eastern extremity of the Lake of Geneva, with sites at various altitudes above it; Méran † in the Tyrol, Lugano ‡ in Switzerland, with the places of different aspect and altitude which surround them, fulfil in a more or less satisfactory manner the requisite conditions.

It would seem at first sight that Pisa § by the side of the Apennine mountains, Pau ‖ with the valleys of the Pyrenees in

---

* See note, p. 243.

† Méran is an Austrian town in the Tyrol, upon the Italian side of the Alps. See "Handbook for Travellers in Southern Germany and Austria" (J. Murray); "Curative Effects of Baths and Waters," Julius Braun; "Handbook for Travellers in the Eastern Alps," K. Baedeker; "European Guide-book for English-speaking Travellers," Appleton; etc.

‡ Lugano is a Swiss town in the canton Ticino, on a curve of Lake Lugano. See "Dictionary of Watering-Places" (L. Upcott Gill); "Curative Effects of Baths and Waters," Julius Braun; "Health Resorts," J. Burney Yeo, M.D.; "Handbook for Travellers in Switzerland" (J. Murray); "Health Resorts for Tropical Invalids," Moore; "Cities of Italy," Augustus J. C. Hare; "Wintering in the Riviera," W. Miller; "Handbook for Travellers in Switzerland," K. Baedeker; "Hachette's Diamond Guide to Switzerland;" "The Alps and how to see them," J. E. Muddock; "European Guide-book," Appleton; etc.

§ Pisa is a town in the north-west of Italy, upon the river Arno. See "Curative Effects of Baths and Waters," Julius Braun; "Health Resorts," J. Burney Yeo, M.D.; "Baths and Wells of Europe," J. Macpherson, M.D. "Principal Southern and Swiss Health Resorts," W. Marcet, M.D.; "Wintering in the Riviera," W. Miller; "Cities of Italy," Augustus J. C. Hare; "Handbook to the Mediterranean" (J. Murray); "Handbook for Travellers in Northern Italy," K. Baedeker; "The Climate of the South of France," etc., C. J. Williams; "European Guide-book," Appleton; etc.

‖ See note, p. 309.

its immediate neighbourhood; the Riviera on the shore of the Mediterranean Sea, with stations on the Alps at no great distance therefrom, offer in the same way every facility for a prolonged abode. Theoretically this is undeniable, but when the scheme is put into actual practice it is found to be quite the reverse. For several reasons the residence at these stations cannot be prolonged until the end of spring, and when they are left, the places in which the patients should reside during the hot season are not yet prepared to receive them. This hiatus in the succession of residences takes away from them the advantage which they might otherwise be as annual residences—a condition of things which might certainly be improved, as without doubt will some day be done, but which at the present time prevents them from being equal to the preceding as regards my plan of fixed abode.

When, therefore, the individual character of the patient and form of the disease enable the privileged stations to be inhabited which, either themselves or conjoined with places in the immediate neighbourhood, can be occupied during the whole year, this plan should certainly be adopted—an advantage, in my opinion, of the highest importance. The preceding enumeration shows that, while making my limit dependent upon this condition, the most dissimilar regions can still be selected, answering to very different indications. Numerous cases thus exist in which the application of this principle meets with no serious difficulty when the second group of climates must be brought into use.

Numerous cases undoubtedly remain in which the condition of the patient imposes residence in a station of the plain which does not present the same resources with regard to the hot season. The question may lie, for example, between Algiers,[*] Egypt,[†] and Sicily.[‡] In these circumstances the idea of pro-

---

[*] See note, p. 297.   [†] Ibid.
[‡] Sicily, the island in the Mediterranean Sea between Italy and Africa

longed residence must be abandoned for the two following reasons. In the case of patients whose condition imposes residence in one of these countries, the indication is so important that all else must yield to it. In the second place, means exist by which this difficulty can be overcome; thus, as regards the second group of climates which includes all the so-called winter stations, an infringement of the rule that the place should be continuously inhabited does not produce by any means the same inconvenience as with climates of the first group, since as respects stations of the plain it is not the most effective element of treatment either for good or bad which is considered, namely, the diminution of atmospheric pressure. Thus it is possible and even easy to find a summer abode which continues with little difference the climatic action of the winter residence; and the change imposed by the hot season is without harmful influence, since the rule laid down is broken in appearance rather than reality. Thus it is that in these special conditions, if precautions are taken to prevent any sudden change of temperature, the rarefaction of the air need not be considered; and, without any compromise or violent rupture of the accommodation established, the indirect application of my principle may be combined with the interest of patients who must pass the winter in southern stations which cannot be inhabited as a summer residence.

It should also be said that this principle need not always be applied with the same rigour when the other stations of the plain are to be employed. An important distinction should be made, based upon the ordinary residence of the patients. Those who belong to a temperate region might

contains some places, Palermo, Catania, etc., which are used by invalids as winter residences. See "Curative Effects of Baths and Waters," Julius Braun; "Handbook for Travellers in the Mediterranean" (J. Murray); "Handbook for Travellers in Sicily" (J. Murray); "Influence of Climate on Pulmonary Consumption," C. T. Williams; etc.

pass the summer and beginning of the autumn in their own country, but upon their return home, after spending the winter at one of the most southern stations of this group, should be careful to remain, during the months of April and May, in some intermediate region. When this is done no real harm can result from adopting the above plan. Should the time have been reached when high climates are inappropriate, nothing can be gained by employing any but those which are merely preservative, and it is for this reason that the possible advantage of abandoning the winter residence in spring will not justify the dangerous suspension of treatment which has a really active effect.

In the case of patients whose residence is in a northern country, the return to their home, after a period of six months, is a serious inconvenience on account of the great contrast in the meteorological conditions of the two regions. When it is once thought necessary in such a case that the winter should be spent at one of the places belonging to the group of low stations, the patients should remain during the whole year under the same climatic influence, lest the advantage of such residence be entirely lost, while all the pre-existing complications are but too often rapidly aggravated.

The question is quite different in the case of patients who inhabit the southern countries of our hemisphere. It is the end of spring, the summer, and beginning of autumn which patients should pass in the temperate regions of Europe, and observation does not show any inconvenience to occur on account of their return home during the winter season. With respect to the inhabitants of the other hemisphere, the reverse period of the seasons should not be forgotten. In such regions the hot season, which would be injurious on account of the high temperature, corresponds to our winter. It is, therefore, at this time that patients should live in this country, which should be left towards the end of spring if such be their desire. It will then be the cool season in those regions, so that no pronounced interruption occurs in the climatic conditions which they have experienced since leaving home. Inhabitants of the tropical regions should spend some years away from home on account of the slight difference which exists between the summer and winter temperature in their own country.

Before leaving this subject, a practice should be mentioned which has become more or less established during the last few years, and which, in my opinion, is most reprehensible. It consists in sending patients to pass the winter at a station in the valley, as, for example, at one of the places in the Riviera * on the shore of the Mediterranean Sea, while the summer is spent at one of the highest places in the group of such rigorous climates as that of Samaden,† St. Moritz,‡ or Davos.§ This practice, which is most prejudicial to the patients, is based upon an erroneous conception as to the mode of action of these two varieties of climate. It embraces, as already shown, two plans of treatment which are entirely different, or, as may be more truly said, opposed to each other, and of which no combination is possible. Unfortunately, one of these two methods of treatment should succeed the other in far too many cases, but they should never be employed alternately. When once the period during which high climates are beneficial has passed, when once the patient is obliged to reside in the milder climates of the plain, this change should be final so far as the extreme stations of the first group are concerned. No return to these places is now possible either in summer or winter, since

* The Riviera (river valley) is the name given to the valley of the Ticino, which, when the river has been joined by the Brenno, increases in size and takes this name. Cannes, Nice, Monaco, Mentone, and San Remo are situated therein. See " Health Resorts," J. Burney Yeo, M.D.; " Principal Southern and Swiss Health Resorts," Wm. Marcet, M.D.; " Winter and Spring on the Shores of the Mediterranean," J. H. Bennet, M.D.; " The Riviera," E. J. Sparks, M.A.; " Influence of Climate on Pulmonary Consumption," C. J. Williams; etc.

† See note, p. 318.   ‡ See note, p. 250.   § See note, p. 243.

patients who have reached what may be termed the second stage of climatic treatment cannot endure without certain disadvantage, without the occurrence of more or less immediate and serious complications, such radical and repeated modification in the condition of hæmatosis, and in the respiratory and circulatory mechanism. Nor can they even be kept during the whole year at the place where they spend the winter, as has been already explained, but the change which is requisite in summer, on account of the excessive heat at the winter residence, should be confined within narrow limits, so that a reasonable change is made, but not one which is so pronounced that gradual accommodation to the climate would be necessary. In my opinion a change of from 1950 to 3300 feet (*de* 600 *à* 1000 *mètres*) is the greatest that should be allowed, the transition from one height to the other being gradually effected. Nor is even this alteration unreservedly admissible in the case of patients who have passed the winter at a seaside station, while on the contrary it is most appropriate, and may even be somewhat exaggerated, in the case of those who have been during that season at a highly situated place such as Méran,* or the region of Montreux.† By whatever motive it may be inspired, a question into which I will not enter, the practice to which I have alluded is certainly to be condemned.

The second question will now be considered, namely, what conditions a station should fulfil that it may be beneficial to patients either liable to be or actually affected by pulmonary phthisis.

* See note, p. 320.

† Montreux is a Swiss village in the canton Vaud, at the eastern end of the Lake of Geneva. See "Dictionary of Watering-Places," part ii. (L. Upcott Gill); "Curative Effects of Baths and Waters," Julius Braun; "Health Resorts," J. Burney Yeo, M.D.; "The Baths and Wells of Europe," J. Macpherson, M.D.; "Principal Southern and Swiss Health Resorts," Wm. Marcet, M.D.; "Wintering in the Riviera," W. Miller; "Handbook for Travellers in Switzerland" (J. Murray); "Handbook for Travellers in Switzerland," K. Baedeker; "European Guide-book," etc., Appleton; etc.

To whichever group the place may belong, the climatic conditions are but one of the elements required to confer upon it the value of a medical station. This is perhaps the most important element in the case, but its consideration will not always suffice, and the neglect of other circumstances connected with the place may expose the patient to serious inconveniences, and bring discredit upon the physician, who, without sufficient knowledge, has recommended a station of which the climate is the only advantageous condition. It is with respect to this point that the advantage of an opinion, founded upon direct and personal acquaintance with the places, is specially obvious. As was said at the beginning of this work, by carefully studying what has been written, a tolerably exact idea of certain climatic conditions of the region may be acquired. No other considerations, however, though equally fundamental and necessary, can be thus obtained, and the physician who wishes to have a real and complete knowledge of the place has no other resource but to obtain it personally.

The conditions which it is indispensable to consider, besides that of climate, are numerous and of different kinds. The principal of these will be mentioned, but it would be impossible to consider them completely without entering into details and minutiæ which are incompatible with this mode of writing.

In the first place, the sanitary conditions of the region must be ascertained. However appropriate a place may be on account of its climate, it should be at once removed from the list of medical stations, if liable to any important endemic complaint, such as malaria, enteric fever, or if frequently affected by severe epidemics, such as diphtheria. The same should be done if the patient can only become accustomed to the country after repeated or, it may be, severe attacks of gastro-intestinal catarrh. The legitimate character of these exceptions will be at once admitted, and the places to which they refer should be known to all physicians; it is unnecessary to dwell upon these facts, and it need only be remembered that they are of the greatest importance with regard to southern stations.

In the second place, the hygienic conditions must be seriously considered. The least that can be expected in a residence destined for invalids is that the ordinary hygienic rules be strictly observed. It should, therefore, be carefully ascertained that the water employed for drinking purposes is potable, and has no communication either with cemeteries or drains. The system of subterranean pipes should be accurately known, and the character of any emanations which might taint the atmosphere, etc., no stations which fail to satisfy the minimum of requirements in a *satisfactory* manner with respect to these points being employed. The term "*satisfactory*," and not "*complete*," is necessarily used, since, with the exception of Madeira,* no station in the second group, that is to say no southern station, seems irreproachable from these different points of view; consequently, if any of them are to be employed, the physician must content himself with an approach to perfection, being careful to choose the places which are least defective in these matters. It is a deplorable fact that the local administrative powers, though constantly warned by physicians, show so little anxiety to correct these defects, which in certain regions of Spain and of the Riviera, both on the French and Italian side, are so pronounced as to be dwelt upon even in the newspapers.

Besides these necessities connected with ordinary hygiene, the places which claim to be regarded as medical stations should fulfil a certain number of special conditions which experience has shown to be of equal importance. The residences should not be placed in a large town, and if the climatic advantages are so great as to justify an exception to this rule, it is indispensable that the houses occupied by patients

* See note, p. 243.

should be far removed from the thickly populated districts, or, still better, be placed at the outskirts of the town in the most favourable position as regards ventilation and purity of the atmosphere.

This has been most happily effected in the French colony of Algiers.* The medical station is not in the town of Algiers, which at present is unsuited to the residence of patients on account of the bad system of drainage and water supply. At some distance therefrom, upon the beautiful declivities of Mustafa Supérieur, a district really adapted to patients will be found, which may be utilized without danger, the imperfections in the town drainage having no effect in these parts. This place, in fact, is situated in the country, having, so far as its daily needs are concerned, the advantage of being in the vicinity of a large town, and resembling in that respect the environs of Cairo,† Palermo,‡ or Catania.§ It need scarcely be added that stations in the country, which are similar in other respects, should always be preferred to those in a town, whether it be great or small.

The houses destined to receive patients should have certain special characters. It is in my opinion a mistake to suppose that the same arrangements are necessary as in the so-termed first-class hotels. These must certainly exist, but even more is requisite. Without dwelling upon the proper aspect of the

---

* See note, p. 297. † Ibid.

‡ Palermo is a town on the north coast of Sicily, towards its western extremity. See "Curative Effects of Baths and Waters," Julius Braun; "Cities of Southern Italy and Sicily," Augustus J. C. Hare; "Principal Southern and Swiss Health Resorts," W. Marcet, M.D.; "Handbook for Travellers in Sicily" (J. Murray); "Winter and Spring on the Shores of the Mediterranean," J. H. Bennet, M.D.; "Influence of Climate on Pulmonary Consumption," C. J. Williams; "European Guide-book," etc., Appleton; etc.

§ Catania is a winter station on the east coast of Sicily. See "Cities of Southern Italy and Sicily," Augustus J. C. Hare; "Curative Effects of Baths and Waters," Julius Braun; "Handbook for Travellers in Sicily" (J. Murray); "Handbook for Travellers to the Mediterranean" (J. Murray); "European Guide-book," etc., Appleton; etc.

house, sufficient cubic size of the rooms, the management and disinfection of sinks, etc., the regular disinfection of linen and bedding which are naturally expected, the necessity of a ventilating apparatus in each chamber, which acts independently of the patient's will, of a heating apparatus provided, if possible, with a chimney in any country except Egypt * and Madeira,† and of many bath-rooms in the house itself, are necessary. Nor is it less advisable that public walks should exist in the neighbourhood, which can be reached by sheltered approaches, and are themselves under cover, being at the same time well ventilated, and having the best position which the country affords both as regards protection from the wind and length of exposure to the sun. Strictly speaking, it is only possible to dispense with these requirements in Egypt owing to the scarcity of rain in that country. A house which does not fulfil them may perhaps be an excellent hotel, but is certainly ill adapted as a residence for the class of patients which is now being considered.

While the minimum of hygienic conditions which should be thought indispensable is thus fixed, in hopes that they may ultimately be realized, it should be added that in this question of special, as in the previous one of general hygiene, more than a relative fulfiment of the different requirements cannot be obtained. The least harmful rather than the most favourable place must be sought, since, notwithstanding the superiority of Madeira, and certain places in Mustafa Supérieur,‡ there is, in my opinion, no place except the mountain sanatoria of the first group which completely fulfils in this respect the requirements which have just been mentioned. These establishments should long have been taken as models by stations of the plain, in which the arrangements, though far more easy to make, present even at the present time notable desiderata.

Nor is this all, since independently of the sanitary contitions, and the general or special hygienic arrangements, the

* See note, p. 297.   † See note, p. 243.   ‡ See note, p. 297.

therapeutic resources of the place must be taken into consideration. It is by no means enough that good physicians and chemists inhabit it. Any place aspiring to the title of medical station adapted to the treatment of phthisis should afford means of treatment by the milk of the cow, goat, or ass taken in the shed, and should be provided with a complete hydropathic establishment, in which the temperature of the water can be elevated or reduced. The varieties of apparatus required for the employment of aërotherapeutic treatment should exist there, and if the fixed pneumatic cabinets which it is costly to install are absent, transportable machines with double action should be obtainable. In places of which the height exceeds 3950 feet (1200 *mètres*), the latter requirement is unnecessary, since the rarefaction of the air at that altitude ensures and maintains the exercise of the lungs far more effectively than machines of any kind. Although in stations of the plain more than one defect still exists in the therapeutic resources—and they can scarcely claim to be faultless in this respect—it must be acknowledged that it is in these places that the most notable progress has been made during the last few years, specially as respects the treatment by milk and the transportable aërotherapeutic machines. Arrangements for treatment by hydropathy have not, unfortunately, shared in this improvement, and the means of employing it are either absent or to some extent useless in many important regions.

At the risk of exciting surprise, to my own discredit, it may be said, as I think, without hesitation, that the question of obtaining food should be considered with the same care, namely, as respects its quality rather than its quantity. This is most important in relation to the more southern stations, and some places in the Canary Islands * and Morocco † will be mentioned, which, if the climate is alone considered, are irreproachable, and which, in fact, have been recommended on

* See note, p. 297. † Ibid.

this account by English physicians, but where, in my opinion, no patient should be sent at present on account of the defective quality of the food.

These, then, are the different conditions which a true medical station should fulfil, nor would it be difficult to enumerate the places which do not deserve this qualification from completely violating one or other of the fundamental conditions. It would not be agreeable, however, to make such an unpleasant statement, and the more courteous plan of leaving such stations unmentioned will therefore be adopted. Thus, if any place usually included in the list of stations adapted to the treatment of phthisis is found to be absent from this work, the cause will not be forgetfulness or personal ill will, the omission being entirely due to knowledge which has been personally acquired—knowledge of inconveniences which cannot be tolerated either from a sanitary or hygienic point of view, or are connected with the material comforts of the patient.

The practical application of the general principles which have been stated in connection with climatic treatment will now be considered. When the prophylaxis of the complaint is in question, or the confirmed disease, unaccompanied by any of the counter-indications mentioned, the stations belonging to the first group should be employed. The principal of these stations in Europe are the following, in order of increasing altitude: Falkenstein * in the Taunus, at an altitude of 1640 feet (500 *mètres*); Görbersdorf† in Silesia, 1830 feet (557 *mètres*); Aussée‡ in Styria, 2300 feet (700 *mètres*); Gaudal§

---

* Falkenstein is a village and tower in the Taunus, a region in South-West Germany. See "Curative Effects of Baths and Waters," Julius Braun; "Health Resorts," J. Burney Yeo, M.D.; "Handbook for Travellers on the Rhine," K. Baedeker; "Handbook for Travellers in Southern Germany and Austria" (J. Murray); etc.

† See note, p. 288. ‡ Ibid.

§ Gaudal (Gaundal) is the name given to a road, hotel, and pension in Norway, 3000 feet above the level of the sea, and upon the road from Christiania to Molde. See "Handbook for Travellers in Norway," K. Baedeker; etc.

in Norway, 2640 feet (805 *mètres*); Davos Platz* in Switzerland, 5100 feet (1556 *mètres*); Samaden † and St. Moritz ‡ in Switzerland, in the Upper Engadine, 5720 and 6090 feet (1745 *et* 1855 *mètres*). On account of their slight altitude, the stations of Falkenstein and Aussée are beneath the minimum limit which was assigned to the first group of climates in the latitude of Central Europe; their management, arrangements, and winter climate, however, cause them to resemble other places in this group, so that they should be looked upon as intermediate between the first and second group. As regards Görbersdorf,§ its more northern latitude compensates for the slight altitude, while its height exceeds the lower limit of stations in the first class which belong to the northern latitudes.

Such are the means which may be adopted during the first stage of climatic treatment, and whenever the choice of places is quite unlimited, the preference should, in my opinion, be given to Davos or the Engadine.

This preference does not imply any inferiority in the arrangements at other places of the group. It may be said that all are model stations, which may be looked upon as irreproachable from the points of view already mentioned. It is but right to say that at Görbersdorf the covered walks and colonnades are of exceptional length, and that it possesses a large winter garden, maintained at a suitable temperature, which is a valuable resource in wet weather. The preference mentioned above is based upon considerations of a totally different nature.

It is a mistake, which seems generally made, to look upon all the preceding stations as of identically the same value, and to regard them as similar results of the doctrine of Brehmer. The first impulse came, certainly, from him. In signalizing, as Graves had done, from the time of his first writings in 1859,

---

* See note, p. 243.   † See note, p. 318.   ‡ See note, p. 250.
§ See p. 331, and note, p. 288.

the benefits derived from life in the open air, an invigorating climate, and the reasonable employment of hydrotherapy; in dwelling upon the importance of exercising the lung, and the necessity of establishing sanatoria above the limits of altitude for phthisis; in showing by precept, and by the example set at Görbersdorf, that the treatment should be methodical—that patients, instead of being free, should submit to the supervision and repeated prescriptions of the physician, not only from a therapeutic but also from a hygienic point of view, being thus as it were at a hospital in these respects; lastly, in opposing the vulgar prejudice which relates to the special advantage of hot places, and proving that the efficacy of methodical treatment is increased by the action of the most different climates,—Brehmer introduced a reform which was not less radical than advantageous in the treatment of pulmonary phthisis. These, I believe, were his own suggestions, for which he deserves the highest honour.

The theory, however, that rigorous climates should be deliberately employed as the most important therapeutic agent, and that mountainous climates are beneficial, even in winter, is not contained, I believe, in the first works of Brehmer, and in my opinion the honour of this discovery belongs to my friend Küchenmeister, and the merit of its first application to two eminent Swiss physicians, Spengler and Ungern. That they were guided by the theory of Brehmer is certain, since Ungern had resided at the establishment of Görbersdorf before going to Davos; but whilst adopting the views and practice of the Silesian physician, they added something to the doctrine which he promulgated. This something, it may be repeated, is the principle, firstly, that some climates have a more beneficial effect than others; secondly, that these climates may be beneficial even in winter—an idea in support of which Spengler wrote his first work in 1862. When, at a later date, I studied this subject myself, I was able, in my turn, to add an interesting proposition to those already promulgated, by showing that when the invigorating effect of climates at a certain altitude is considered, the latter is but one cause of their decided utility, the rarefaction of the air exercising a therapeutic effect of equal if not greater importance.

It will now be understood why, though recognizing how perfect the arrangements are at other stations of the group, I prefer Davos * or the Engadine;† namely, because the climatic conditions, which are in my opinion of such utility, are far more completely realized in those places. As regards, in fact, the fundamental element of utility, the rarefaction of the air, no comparison can be made between Davos or the Engadine and other places, the altitude of the Swiss stations being 5110 feet (1556 *mètres*) in the case of Davos, 5720 feet (1743 *mètres*) in that of Samaden,‡ and 6090 feet (1855 *mètres*) at St. Moritz;§ that is to say, twice or three times as much as at the other places. On the other hand, while offering this fundamental advantage, which, in my opinion, is the most important consideration, Davos and the Engadine are equally well furnished with certain climatic properties of incontestable value, which cannot be obtained elsewhere to the same extent.

The rarefaction of the air shown to exist at Davos by the mean barometric height of 24·6 ‖ (626 *mm.*), and in the Engadine of 24·2 inches (616 *mm.*), is not the sole effect of the altitude in these regions; the association of high altitude with a comparatively southern latitude causes a special dryness of the air, a circumstance which is most favourable owing to its coincidence with a low temperature. The cold is on this account far better tolerated, and the tonic influence which it has upon the organism produces more effect from not being

---

* See note, p. 243.   † Ibid.   ‡ See note, p. 318.
§ See note, p. 250.
‖ It will be remembered that, speaking roughly, a depression of an inch in the height of the barometer is produced by an ascent of 900 feet, its mean height at the level of the sea being 29·96 inches.

weakened by the opposing influence of humidity. These conditions also produce an atmosphere of unequalled purity, and even without special examination vibrios and other low organisms may be assumed to be absent, since simple exposure to the air is the process adopted for drying meat, which can then be preserved for an indefinite time without change. The incomparable brightness of light, the vividness of blue colour in the sky, the complete absence of fogs even in winter, the intensity of solar radiation, the remarkable frequence of fine days, are other no less interesting effects of altitude combined with a southern latitude.

The benefits resulting from this association are also increased by the position of the places. The protection from northern winds is in both cases complete, the establishments being exposed to the south, and the configuration of the adjacent mountains such that the daily action of the sun is of unexpected duration. Even on the shortest days of the year, at the end of December, the sun shines throughout the valley of Davos from half-past nine in the morning until half-past three in the afternoon. This I was able to prove on the 26th and 27th of December, 1880. At Samaden and St. Moritz, on account of their higher altitude, the sunlight is of even longer duration, exceeding that at Davos by half an hour, a quarter of an hour in the morning and the same in the afternoon. This I found to be the case at almost the same time, namely, from the 29th to the 31st of December. In the spring, summer, and autumn, winds exist which are special to the two regions, and of which the time and direction are almost constant; during winter, however, there is an absence of wind, which must be looked upon as an invaluable advantage, and this circumstance, combined with the dryness of the air, diminishes to an extent which is truly incredible the impression produced by cold.

In addition to this, on account of the long duration of the sunlight by day, and the large number of fine days in winter, the temperature is not so low as would at first be thought. At Davos during the medical day, that is to say the time during which patients can be in the open air, which may be taken as lasting during the six months of winter from ten o'clock in the morning until three in the afternoon, or from nine until four, according to the season, the temperature is usually above freezing point.

Again, by means of works written by the physicians and officers of the sanatorium at Davos, the mean temperature of the three coldest months of the year, namely, December, January, and February, at ten, one, and three o'clock repectively, can be ascertained. In this way the mean temperature of the three months is found to be 35·6° F. (2·03° C.) at ten, 43° F. (6·16° C.) at one, and 40° F. (4·44° C.) at three o'clock.

The temperatures taken as the basis of this statement were obtained by means of a thermometer, suspended freely in the open air and isolated from all surrounding objects. Placed at the extremity of a terrace, it was beyond the reach of rays radiating from the walls of the house. Upon fine days, however, it was necessarily exposed to the sun's rays, since at the times mentioned no point whatever of the valley is in the shade. Even supposing that the above-mentioned figures apply to a series of years which were specially favourable—as there is no reason to believe—the difference is still great between this reassuring fact and the false or interested suppositions due to preconceived ideas or designed partiality.

These notions will be completed by the account of some observations which I myself made with the greatest care.

On the 27th of December, 1880, at seven o'clock in the morning, I observed the thermometer in the open air to stand at 25° F. ($-$ 4° C.); upon the same day at eleven o'clock, by the side of the ground consecrated to skating, the same thermometer marked 46° F. ($+$ 8° C.).

On the 28th of December, at the moment of leaving the valley, I observed it to stand at 34·6° F. (1·5° C.) at eight o'clock in the morning, and at midday, upon the summit of the Fluela Pass, at the height of 7895 feet (2405 *mètres*), that is to say, 2785 feet (900 *mètres*) higher than Davos, the temperature was 33° F. (+ 0·5° C.) Two hours later the temperature at Suss, in the Lower Engadine, at an altitude of 4700 feet (1431 *mètres*), was found to be 37° F. (+ 3° C.), which at this time of day would correspond with the temperature of 34·6° F. (+ 1·5° C.) observed at Davos in the early morning.

It will have been perceived, therefore, that, owing to exceptional circumstances, the reduction of temperature during the medical day presents nothing to cause anxiety or even disquiet. In the Upper Engadine, at Samaden and St. Moritz, the conditions of temperature are most analogous, the difference in the mean monthly temperature of the two districts rarely exceeding 2·6° or 3·6° F. (1·5° or 2° C.), that of Davos being sometimes the highest, sometimes that of the Engadine. There is nothing certain with respect to this, and the two valleys may be looked upon as in all respects analogous as regards the climatic conditions of winter.

On the 29th of December the temperature at Samaden was found to be 36° F. (+ 2° C.) at nine o'clock in the morning, and upon its northern side, being at the same moment 32° F. (0° C.) at Davos. At ten o'clock on the same day it was 36° F. (+ 2° C.) at St. Moritz. On the 30th I observed the temperature at Samaden, in the same place and at the same hour as on the previous day, to be 34° F. (+ 1° C.), the temperature at Davos being at the same moment 36·5° F. (+ 2·5° C.).

On the 31st I found the temperature to be 21° F. (− 6° C.) at nine o'clock in the morning before the sun appeared, the temperature at Davos being 29·5° F. (− 1·4° C.). At one o'clock, when the sun was shining brilliantly, the temperature, as shown by the thermometer suspended freely in the open air, was 30·2° F. (− 1° C.), that at Davos being the same at this moment.

It should be added, in order that nothing of importance may be omitted in the comparison of these two districts, which is so interesting from a climatic point of view, that on account of its higher altitude the dryness of the air is more pronounced and constant in the Upper Engadine than at Davos.

The power of solar radiation, one of the most striking characters in the climatology of these two districts, has been repeatedly mentioned. An observation, artificial in character perhaps, but instructive, will give an idea of this remarkable fact.

On the 26th of December, 1880, in Davos, at nine o'clock in the morning, the thermometer in the open air, and separated from every surrounding influence, marked 16° F. (— 9° C.); being then placed against a wall and fully exposed to the sun, it rose in half an hour to 59° F. (+15° C.).

On the 31st of December, at Samaden, the thermometer, placed freely in the open air at nine o'clock in the morning, marked 23° F. (— 5° C.); the instrument, being then exposed to the sun as at Davos, rose within half an hour to 69° F. (+ 20·5° C.), and at one o'clock on the same day to 86° F. (+ 30° C.).

Upon the same day the temperature in the full sunlight reached the height of 91° F. (+ 33° C.), having risen two days previously, that is, on the 29th, to the height of 106·6° F. (+ 41·5° C.).

Such numbers as these are exceptional, but a temperature of from 77° to 86° F. (*de* 25° *à* 30° C.) is usual in the two districts as expressing the greatest heat of the day in places exposed to the sun. It will thus be understood that patients who are warmly clothed can remain for a long time seated in such parts, and that parasols are the necessary accompaniment of furs—facts which I again verified in my recent journey to those parts.

This fact has another consequence of no less importance: from eleven until three o'clock in the afternoon the windows can remain fully open, and allow the sun's rays to penetrate into the room, which still continues to be heated from within.

This condition, which is indispensable that apartments may be healthy, and which is certainly preferable to the plan of ventilating by means of apertures or machines, can thus be as well realized at high altitudes as in southern countries. Mechanical ventilation is agian assured at every station, being independent of the resident's will, so that life indoors is attended by the most satisfactory hygienic conditions, in the same way that life in the open air exercises the most healthful climatic influence.

The shortness of the time during which patients can enjoy by day the benefit of out-of-door life has been regarded as an objection to their wintering in these places. This objection seems to be unreasonable. It is only from the 15th of December to the 15th of January that this period is reduced to a minimum, and even at that time it lasts from ten in the morning until three in the afternoon, a fact which I have myself found to be true. Before the 15th of December and after the 15th of January the length of time which can be spent in the open air increases by half an hour at each end of the day. It soon includes the whole day from nine o'clock until four, the condition of things not differing in this respect from what occurs in the southern countries. Even there it would be imprudent to allow patients to go out of doors before ten o'clock in the morning, or remain out later than three in the afternoon between the 15th of December and the 15th of January. Thus no serious difference exists between the stations of these two different regions.

An objection has also been made by Richter since the year 1870, which seems to carry some weight, namely, that it is inconvenient and dangerous to unite a large number of patients suffering from phthisis in the same establishment. This, however, is not in any way peculiar to the sanatoria placed at a high altitude. An impartial observer will acknowledge that what takes place at the hotels of the southern stations is far less satisfactory. In addition to those suffering from phthisis, numerous others reside in these hotels affected by entirely different complaints, so that it is not the bad effect which the more severe cases of phthisis may have upon those who are less seriously affected which is alone to be feared, but, what is far more serious, the possible influence which patients who suffer from this complaint may have upon those who do not. Even if this unfortunate possibility is disregarded, and patients who already are or may be affected with phthisis are alone considered, the inconvenience is still more pronounced in stations of the plain than in those at a high altitude. The reason of this fact is that the air is naturally less pure in stations of the valley, that it does not prevent the development of low organisms, and that in the establishments there is neither the plan of independent ventilation which exists at the sanatoria of highly situated stations, nor that of methodical disinfection as regards sinks, linen, bedding, in short, all the articles used by the patients, a system which has been for a long time strictly carried out at the principal high stations. It is true that this argument is founded upon the character of the establishments, but at the same time to the special disadvantage of stations in the plain. In these latter the inconveniences might be avoided by the use of a private house instead of an hotel. The same possibility, however, exists at Davos, so that this objection is of no importance.

Lastly, the objection has been made with a fear, perhaps more apparent than real, that the cold air may produce an impression upon the bronchial tubes and lungs, and that inflammatory affections may be the consequence.

This objection is proved to be unreal by the observations of nearly twenty years, and by the experiments of Heidenhain, which, though made for a different reason, clearly prove that the temperature of inspired air, whether low or high, has no effect upon the lungs so long as it is dry. It will not be forgotten that dryness is one of the most characteristic properties of the air both at Davos and in the Engadine. The necessary consequence of this dryness, combined with the rarefaction of the air, is the abstraction of so much water from the respiratory tube that the lungs are proportionately colder, the consequence of which may be that some antipyretic effect is produced. It is certain that on account of the low temperature of the external air its dryness must be beneficial in all respects without presenting any of the inconveniences and even dangers that it might produce if the temperature was elevated to a corresponding degree.

Having analyzed these objections, which are somewhat slight and unfounded, but a few words need be added that the climate and winter season at these two extreme stations may be completely understood. At Davos, as in the Engadine, there are but few days upon which the weather is unseasonable. It is more favourable during the six months from October to March than in the other half of the year, the days of bad weather invariably resembling each other. A more or less abundant fall of snow occurs, which is scarcely over when the sun and blue sky reappear without the long intervention of a cloudy sky. The wet weather, again, often occurs at night. When the snow has once ceased to fall, heavily charged sledges are put in motion and compress it in such a way that within a short time the roads and paths are covered by a bed of snow, compact, hard, and dry, and which, without being slippery, leaves neither a flake of snow nor drop of moisture on the feet. Another interesting peculiarity about the wet days is their isolation. Never occurring for days consecutively, it is but rare that two succeeding days are wet, so that confinement within doors is never of long duration. Nor is such confinement absolute, since covered promenades and galleries under glass exist which can be utilized by patients both at Davos and in the Engadine.

Another peculiarity must be mentioned, not as yet signalized, but which is of great interest. In these regions the temporary cessation of out-of-door life has not the same consequence as in stations which are more lowly situated. In the latter, confinement to the house involves the total cessation of whatever favourable influence the climate might exercise, while at Davos and in the Engadine this effect is but partial. The tonic action of the air, its good effect upon the nutrition, are interrupted, but the mechanical effect produced by its rarefaction remains unchanged, and the treatment, so far as its fundamental action is concerned, continues to exercise a salutary influence without any time being therefore lost.

Upon fine days, independently of methodical walks upon ascending ground, the patients have the opportunity of taking exercise in a manner which is both useful and agreeable. In the first place there is skating, a vast extent of ground being carefully kept in order for this purpose at Davos, Samaden, and St. Moritz; in the next the sledge, worked by hand, forms a diversion when it slides down an appropriate hill, but becomes a powerful means of exercise when drawn to the summit of the slope which it is then again to descend.* Ladies of all ages may be seen to join in this healthful pleasure with an enthusiasm which it is agreeable to see, such life in the open air, which this climate alone permits, being remarkably adapted to the work of constitutional regeneration which it is sought to obtain.

When the spring weather removes the covering of snow from the surface of the valleys, the place has been so thoroughly and repeatedly watered that the inhabitants are preserved from the inconvenience of dust, which the local winds would otherwise produce upon the roads when their force is unusually great.

* Allusion is here made to the Canadian amusement of "Toboggoning."

These details, all of medical importance, being known, the special appropriateness of Davos and the Engadine to patients who are to be treated by high climates will now be understood. Undoubtedly, the other stations of the first group, notably Görbersdorf, Falkenstein, and Aussée, as already observed, are no less irreproachable as sanitary establishments; but their insufficient altitude, the difference in their latitude and situation, deprive them of the special climatic conditions which are characteristic of the two Swiss stations, and to which they owe their unequalled superiority. The unrivalled beauty of these places when their winter mantle glitters in the sunlight, and of which the dazzling majesty defies all description,, should also be mentioned.

It has been shown that, as far as climatic conditions are alone concerned, Davos and the Upper Engadine resemble each other in all respects. Great progress has been made during the last few years at Samaden and St. Moritz as regards the requirements of patients who winter at those places, and it is evident that in a short time the essential advantages of Davos will be there reproduced. The true and notable difference between them is solely the great distance of the Engadine from all railway stations, and the interposition of a mountainous chain, which must be crossed by highly situated passes. When the railway line of the Tyrol is complete, and Milan is connected with Chiavenna, the Engadine will be placed on the same footing as Davos with respect to its distance from the railway termini. These facts again become of far less importance when the treatment by high altitudes is applied, on account of the principle laid down that the patient should become accustomed thereto during the summer or autumn weather, and are only mentioned in order that these considerations may be in no way incomplete.

For the information of those in the medical profession, as well as of patients, this subject may be left for a moment, that the efficacy of spending the winter at Davos or in the Engadine, when other pathological conditions exist, may be mentioned. Chronic pneumonia, persistent pleurisy, an illness of a totally different nature, namely, unceasing dyspepsia, and the neurotic conditions evidently attached to constitutional debility, are of such a kind. As to the two latter groups of cases, every reserve being made as regards indications resulting from personal excitability, the higher altitude of Samaden and St. Moritz seems to be the cause of their superiority.

The descriptions which have been given will enable the distinction made between Davos and the Engadine on the one hand, and the other stations of Europe, in which the winter climate is severe, as typified by Görbersdorf, on the other, to be appreciated. The principles and details of the question would be completely misunderstood if all these places were considered similar, the difference between them being so great that the following proposition will, in my opinion, briefly explain their characters. At Görbersdorf, Aussée, and Falkenstein it is the physician who acts; while at Davos, Samaden, and St. Moritz it is the climate, since their altitude is the principal and characteristic agent. Medicinal treatment can be applied in these places according to the same principles as in others, that is to say, by following the rules laid down by Brehmer; and it is therefore evident that the Swiss stations have a special effect which belongs exclusively to them, and owing to which they are preferable to other places in the same group. This special effect results from the altitude and climate of the stations, and is totally independent of the physician.

The application of this treatment, except during the prophylactic period of the disease, requires much prudence. This must never be forgotten, and great attention should be paid to certain precautions, which are, in my opinion, indispensable.

No patients should be sent to Davos or the Engadine during winter, or even at the end of autumn, if this is their first visit to those places. The most suitable time is either the summer, before the end of September, or the month of October, the equinoctial period being carefully avoided either for the time of the journey or arrival at the station. The patient then becomes acclimatized to the place during the warm weather, and, adapting himself to the new conditions with more ease and rapidity, will before winter have already acquired the advantage of such constitutional improvement as will permit him to pass through the cold season without confinement to the house or danger to himself. The most extreme cold always comes on gradually, and thus the healthful influence of the climate can be continued without interruption or dangerous shock to the patient.

To appreciate the whole importance and necessity of this rule, it should be remembered that the patient who is directed to live at a high altitude must accustom himself not only to the rarefaction of the air which exists at all seasons, but also to the low temperature of the place. It is neither reasonable nor prudent to impose upon him the necessity of making these two changes simultaneously owing to late arrival at the station. The method which I employ has another advantage, which experience has shown me to be of no little value. At the time indicated, sledging has not yet begun, and the patient can reach his destination by carriage. Notwithstanding the improvements made by the Swiss post in the sledge service, there is still much to be desired, and the use of a carriage is certainly to be preferred in the case of those who have already passed the prophylactic period of the disease.

Patients who live usually in the plain should not be allowed to go directly to Davos or the Engadine. It is true that some physicians of these places look upon such a precaution as quite unnecessary, but for my part I cannot share the opinion, for reasons already given. The patient should break the journey at two places, remaining at least five days at each, one at the height of from 1650 to 2000 feet (*de* 500 *à* 600 *mètres*), the other at that of from 3300 to 3600 or 4000 feet (*de* 1000 *à* 1100 *ou* 1200 *mètres*), such a proceeding being absolutely necessary to ensure his becoming accustomed to the place with ease and rapidity.

The preceding rules admit of no exception in the case of patients coming from the temperate or southern regions, representing in these cases the minimum of precautions which are indispensable. In the case of those who were born and have lived in countries which are unmistakably to the north, this healthful severity is less necessary, and Davos may be visited without inconvenience during the autumn or even in winter, the principle of successive stages being always observed, and upon condition that long residence in a temperate or warm climate has not removed their power of adapting themselves to those which are more severe.

Such being the rules imposed by prudence, what should be done in the case of a patient to whom the treatment by high altitudes is considered beneficial or necessary, when the period is too late for a visit to highly situated places? When this happens, as is not rarely the case, the patient should be instructed to pass the winter, which is beginning or has begun at some other station of the first group, Norway being excepted, which is unsuitable as a winter residence. Thus, according to the case, he should be sent to Görbersdorf, where the climate is severe but the altitude slight; or, again, to one of the intermediate stations, such as Falkenstein or Aussée, at which the altitude is high but the climate less severe. There the treatment can be carried on methodically, the arrangements are faultless, and at the end of spring or in the summer Davos or the Engadine may be visited without inconvenience. It would, again, be possible to send the patient in such circumstances to one of the extreme stations in the second group; that is to say, to one which, on account of its climatic conditions in winter, differs but slightly in this respect from the stations at a high altitude, but which, on account of its different height, does not impose the same necessity as to the patient becoming gradually accustomed to a rarefied atmosphere. Méran,* Montreux,† or Lugano ‡ might be utilized in such cases, and at the end of spring or beginning of summer, after fulfilling the rule laid down with respect to successive stages, numerous means of doing which exist in the more or less immediate vicinity of these places, the patient might remain at Davos or in the Engadine, where he would arrive in the best conditions possible, both as regards adaptation to the climate and success of the treatment.

However rational this second plan may appear, it cannot be considered equal to the first, since if the patient could spend the winter with advantage to himself at Méran, Montreux, or Lugano, the same may be said with respect to Görbersdorf,§ Falkenstein,‖ or Aussée.¶ These places would be more beneficial on account of a more invigorating climate being there associated with a higher altitude, and owing to the superiority of the establishments in those places from a medical point of view. Consequently, in the conditions now being considered, namely, the indication of treatment by severe climates and high altitude at a time which forbids employment of the extreme stations of the group, a more lowly situated station in the same group should be inhabited as the best substitute, Méran, Montreux, and Lugano being reserved for cases in which the patient firmly refuses to reside at the above places.

These details might perhaps be considered unnecessarily minute, and the precautions imaginary. This, however, would be a mistake; it is from ignorance of the former and neglect of the latter that the physician is liable to make patients lose time, which is of great value, and to compromise, owing to its

* See note, p. 320.   † See note, p. 325.   ‡ See note, p. 320.
§ See note, p. 288.   ‖ See note, p. 331.   ¶ See note, p. 288.

bad application, treatment which, when properly directed, has been shown by experience to give most satisfactory results. Nothing has been said that might have been omitted; nothing, in fact, which is not indispensable, that this method of treatment may be truly effective.

It is unnecessary to say more as to the general indications and counter-indications of the severe climates at a high altitude, these having already been considered, but it would be as well to mention once more the principal places at which such a climate exists, specially Davos and the Engadine. The details have already been discussed at some length in a previous chapter.

The treatment by prolonged residence at these places is invariably indicated when prevention of the disease need alone be considered. The precautions to be taken, and the gradual transition by means of which these places should be reached, are the more necessary in proportion as the patient is weaker or more excitable.

When the disease exists, there is one, the so-called pneumonic form of the disease, which definitely excludes Davos and all other stations of this group, at any rate in winter.

These places, however, might be employed during the spring and summer, should this form of disease persist without interruption for many months, and the inflammatory foci, being stationary and inactive, show no sign of active congestion at their periphery. In these conditions the effect of the rarefied atmosphere may be favourable to the partial resolution at least of these primary lesions. The decision is of a most delicate nature, but if correct, will undoubtedly be beneficial to the patient. It is, again, a question of individuality, and its solution, defying all rule, can only be obtained by means of observation and experience.

In the ordinary form of tuberculosis it is the character of the reaction which should be the fundamental guide. Extreme altitude is beneficial to patients with torpid and ill-pronounced reaction, while the reverse is the case, whatever the condition of the patient may be, when the individual is of an excitable disposition, and the reaction active, namely, in the case of those who present one of those forms known by the name of florid reaction or erethism.

High climates are also counter-indicated in those cases which are accompanied by fever from the first, as also when the pyrexia occurs later in the disease, if of the remittent type, and continuous, notwithstanding appropriate therapeutic intervention. On the other hand, intermittent pyrexia, in which the attack occurs in the evening, is not in itself a counter-indication.

Hæmoptysis furnishes no indication in either sense, everything depending upon the phenomena which coexist or follow, and upon the type of reaction in the patient. This question has been previously discussed at length, and it will only be now stated that in such a case residence at a high altitude has invariably an active effect, either favourable or unfavourable, according as it corresponds to the indication which exists or not.

In ordinary phthisis, the character of the local lesions furnishes, in my opinion, no indication with regard to residence at a high altitude. Whether catarrh of the apices, foci due to previous pneumonia or broncho-pneumonia, or miliary tubercles exist, there is nothing in the pathology of the disease alone which can influence the decision. The age of the lesions, again, is of no moment. Considered, therefore, from these two points of view, the local alterations in no way prevent the decision being entirely dependent upon the mode of reaction, the pyrexia, and general condition of the patient. On the other hand, the extent of the lesions is of the greatest importance in this respect. Whatever the character or age of the alterations may be, if they are so extensive as to interfere with hæmatosis, and maintain a permanent condition of, it may be, unperceived dyspnœa, residence at even such a high altitude as that of Davos will be counter-indicated by this alone, whatever the other indications may be. If, in these conditions, treatment by the severe climate of winter is considered necessary, the Swiss stations must be abandoned, other places in the group being utilized which are now more precisely indicated. The altitude is not so great, specially at Görbersdorf and Falkenstein, that the diminution of atmospheric pressure need be taken into consideration, and the patient, without fatigue or danger to himself, may derive benefit from the invigorating effect of the climate and the methodical treatment instituted at these places.

For the same reason, namely, from the special effect of rarefied air, affections of the bronchial glands, generalized emphysema, abnormal conditions of the heart or large blood-vessels, definitely preclude residence at the extreme altitude attained by this group. Experience has also shown that laryngeal complications, attended by ulceration, enteritis of an ulcerative character, or nephritis in any form, should be considered positive counter-indications.

This chapter will be ended by a statement which is, in my opinion, of great importance, since no one is more anxious than I am to combine the interest of the patient with the prudence of the physician. If in any case doubt is felt as to how climatic treatment can be organized; if, whilst no doubt is felt as to the propriety of treatment by severe climates, some particularity exists which causes indecision, no step should be taken in this direction. Should the difficulty not be removed by acquaintance with the case in question, this must not be expected from the effect of chance inspiration. In such circumstances, which frequently occur in the course of practice, there is a simple means of arranging everything without danger being incurred. Nothing should be done precipitately, but, the idea of extreme altitude being abandoned, recourse should be had to the lowest stations of the first group; and if this precaution does not seem to meet all dangers, the places which are considered intermediate between those of the mountain and valley may be utilized. The dictates of prudence will be thus respected, and the benefit produced by a strengthening climate be assured to the patient without danger being incurred from the altitude of the station. After the first winter, which, if spent as above mentioned, cannot possibly be injurious, all doubts will be removed, the way will be clearly traced, and if severe climates are indicated as before, it will be safe to utilize in the second year the extreme stations which had as yet been prudently avoided. Such treatment will have been beneficial to the patient in a medical sense, for it should not be forgotten that the most urgent and constant requirement of patients is the beneficial effect of treatment by high altitudes and mountainous climates as often and for as long a time as possible. This method may in all truth be recommended, being one to which I often had recourse when less conversant than I am now with the solution of this therapeutic problem.

This discussion may be summed up in the following short but precise conclusion: Davos, Samaden, St. Moritz, are the stations of fundamental value in treatment by high climates. On account of the great difference in their elevation, the other places in this group, Görbersdorf, Falkenstein, Aussée, may be looked upon as possible substitutes, which should be reserved for cases in which extreme altitude is counter-indicated, though climates in which a cold winter exists are permissible, and for those patients who, from the influence of prejudice or routine, are unwilling to reside in the most elevated stations.

Notwithstanding the clearness and apparent precision of the rules laid down, it should be understood that, except during the prophylactic period, the decision as to residence at a high altitude is one of the most delicate and difficult problems in medical practice. It is a purely individual question, and its solution, settled by no law, may be influenced, but cannot be supplied, by dogmatic rules. The difficulty, however, must not be shunned, as might easily be done by the simple resolution to avoid utilizing high altitudes or severe climates in the treatment of phthisis. Such a proceeding, however, would be unjust to the patient, who would thus be deprived of a resource in therapeutics which is really indicated, and injury would be done, not only by this omission, but also in all probability by the different treatment which is substituted for it. It has already been shown that the above treatment is effective in doing either good or harm. Never inactive, it is beneficial or injurious according as it is more or less appropriate to the individual conditions, and the difficulty in applying it is increased by the fact that observation and experience are the only certain guides to a decision which must vary according to the case. This is true, and cannot be too much insisted upon; but would it be right to abandon on that account the whole method of treatment? Such would be an erroneous conclusion, and a mistaken view of the physician's duties. Is digitalis unemployed owing to the fact that in some cases it may do as much harm as good? By no means. Nor in this case would it be right to seek a convenient refuge in the *statu quo* and by following routine. The general precepts should be considered, the distinctions to be made in following the rules should be understood, the patients themselves should be studied in connection with these laws, and this new therapeutic difficulty may then be met without that fear of error which the serious responsibility of the case is liable to impose upon the physician.

It would, again, be a mistake to suppose that these difficulties and dangers are exclusively connected with the question of treatment by places belonging to the first group. The very same occurs with regard to places in the plain when these are taken into consideration. Without doubt, when stations of which the climatic effect is well known, and but slight in degree, such as Pau,* Pisa,† and the different places ‡ in the Riviera on the shore of the Mediterranean Sea, are called in question, there is but little difficulty and no danger, owing to the small effect produced. When, however, the extreme stations are considered, as they should be, namely, those in which the climate has a more decided and powerful effect, such as Méran § on the one hand, and Madeira,‖ Algiers,¶ Egypt,** and Sicily †† on the other, the same uncertainty and doubt exist as before, and a mistake would be no less prejudicial to the patient. Such being the case, should the inactivity of routine influence the course adopted? What benefit could be derived from expectant treatment? None. Even if the difficulties are numerous, and doubts exist on all sides, these should not be avoided but opposed and overcome.

It is hoped that the details to be now considered will render this a less arduous task.

* See note, p. 309. † See note, p. 320.
‡ Places in the Riviera on the shore of the Mediterranean Sea: Cannes, Nice, Monaco, Mentone, San Remo, etc.
§ See note, p. 320. ‖ See note, p. 243.
¶ See note, p. 297. ** Ibid. †† See note, p. 321.

CHAPTER XIII.

CLIMATIC TREATMENT (*Conclusion*).

General principles—Effect of climates belonging to the second group—Local protection—Constitutional repair.

Climatic conditions requisite to produce these effects—Uniform temperature—Oscillations during the day, upon successive days or months—Wind and dust—Hygrometric condition of the air—Special conditions of the atmosphere—Climatic conditions requisite for constitutional repair—Distinction between exciting and fortifying action—Sedative and debilitating action—Importance of this distinction—Examples—Preponderating value of the fortifying action.

Causal or pathogenic (*pathogéniques*) indications—Symptomatic indications.

New division of climates of the valley—Practical applications—Relation between the pathological and climatic groups—Pneumonic phthisis—Ordinary phthisis—Conclusion.

If the treatment by high altitude and rigorous climates cannot be employed, or must be abandoned on account of some counter-indication, the question arises whether the other phase of climatic treatment should be adopted. What is to be done in such a case? what course would be most beneficial to the patient, and how may the resources which the physician has at his disposal be most judiciously and appropriately utilized?

At first sight the question appears difficult on account of the large number of stations in the second group already enumerated; this difficulty, however, is less serious owing to the fact that many places in that group should not really occupy a place in it. It will be remembered that many conditions besides the question of climate are indispensable at a station designed for patients affected by phthisis, and that the most important character in a climate is its uniformity. It will then be perceived that the long list of places in this group is due either to absence of the necessary severity or insufficient knowledge of the stations, so that the difficulty in deciding between them is more apparent than real. Its solution will be facilitated even more when the stations are mentioned, which, in my opinion, should alone be employed.

Before commencing this discussion, a statement already made with respect to watering-places will be repeated. The course which, in my opinion, should be pursued differs from that employed by my colleagues. In this chapter, however, it is not intended to criticize the practice of others, or to compare it with that which I believe to be the best. The plan which I adopt will be simply mentioned, and reasons be given for its employment, in the sole hope that this knowledge may be useful, and without the existence of any other motive.

Recalling once for all the opinion which I hold, that the residence should be fixed, and the special modes of applying this principle according to the region and home of the patient, some general principles will be first mentioned which are usually misconstrued or ignored, and which, in my opinion, are indispensable guides, and enable the reasons which direct my practice to be clearly understood.

The characteristic features of places in the second group, and the benefits derived from their employment, will not be forgotten. Whilst those in the first group have a direct influence upon the disease, and may even be called curative, or at least be looked upon as important and active remedies, those in the second group are solely and passively preservative in action, and the only effects which may be expected from them, in the absence of any other therapeutic treatment, are the following: diminution of the irritability and consequent irritation of the bronchial tubes, preservation in a greater or less degree from inflammatory complications in the respiratory organs, combination of these effects with life in the open air, and sunlight during a longer or shorter time in winter, and in consequence of these conditions a favourable influence indirectly produced upon the patient's constitution. These effects are undoubtedly of great importance, since, on account of the fortunate combination of a local *statu quo* with improvement in the general condition, the patient is able to derive benefit from every favourable change which takes place in the natural evolution of the disease, or is produced by therapeutic treatment.

I combine and sum up these effects under the two heads of local preservation and improvement of the constitutional condition, or, as it may be briefly expressed, the local and general effect. With the exception of these results, I believe the climates of the plain in the second group to have no action whatever.

Such being the case, the first question to be solved is the following: What climatic conditions are capable of ensuring the most beneficial effect possible? This question, which is new, and has not yet been sufficiently considered, comprises and regulates every practical conclusion. The principal of these conditions will, therefore, be mentioned under the two heads already mentioned.

With respect to local preservation from disease, it might be and often seems really supposed that this specially depends upon the temperature of the place, and that the higher the winter temperature is the more likely is such preservation to occur. It is precisely this opinion which originates the disputes as to the exact degree or even tenth of a degree of temperature in which patients should live, which have but too often been witnessed among physicians of the different stations. The fruitlessness of these discussions has already been mentioned, since the idea is completely erroneous. What is of real importance, as regards preservation from local complications, is not the absolute degree of temperature, but its uniformity, that is to say, the small amount of oscillation in the thermometer. This should be inquired into, both as respects daily and monthly variations. With respect to those which occur during the day, two distinct circumstances should be considered, namely, the changes which take place at different hours of the day, or real diurnal oscillations, on the one hand, and those which occur from day to day on the other. It may, in fact, be said that the uniformity in the temperature of a climate depends upon the three following conditions: difference in the height of the thermometer at different hours of the same day, or from day to day, or month to month.

Such is the question of real and preponderating importance. In proportion as the variation at these different times is smaller, and in consequence the uniformity which exists more complete, to the same extent is the chance of preservation increased and more certain. The extreme extent of variation is but a secondary and accessory fact, and it is the oscillation of the temperature which should be mainly considered; that is to say, a daily oscillation of 9° F. (5° C.) between the extreme temperatures of 40° and 50° F. (*de* 5° *à* 6° C.) constitutes, with regard to the special question of local preservation, a far more beneficial climatic condition than one of 16° F. (9° C.) between the temperatures of 60° and 75° F. (*de* 15° *à* 24° C.), and the same may be said with regard to the two other varieties of oscillation. Thus the uniformity in temperature as above defined is the most important condition required to produce the effects which have been grouped together under the term local preservation.

As to the daily oscillations, properly so called, which are undoubtedly the most important, a difference in the mean temperature which exceeds 9° F. (5° C.) indicates that the fundamental advantage of the climate is compromised, and that it cannot be regarded as of undoubted benefit on account of this change in the temperature. It should be observed that the mean daily oscillations are considered, in consequence of which the extreme numbers in the season of seven months include both higher and lower temperatures than that given. This being understood, the climates in which the daily oscillations of temperature are in the mean greater than the above-mentioned numbers cannot, in my opinion, be considered uniform, and the therapeutic value of the climate is smaller in proportion as the mean temperature is higher than the limits named. Uniformity in the other meteorological characters is not less indispensable, specially as regards the hygrometric condition, since notable variations therein are both distressing and prejudicial to the patients.

Another indication, of no less fundamental importance, is furnished by the frequency and violence of the winds. With respect to local preservation, this circumstance is far more important than elevation of the mean temperature. A region without or with but little wind during the autumn, winter, and spring, and with a mean temperature which is relatively low in the group of climates which we consider, is undoubtedly more beneficial than one in which the mean temperature is higher, but which is frequently exposed to more violent winds. This is true, whatever the direction of the wind may be; and though, undoubtedly, a north wind is the most prejudicial, still any violent displacement of the air is a real obstacle to the preservation desired, since, for the purpose of avoiding this danger, it may be necessary to confine the patient to the house. . Independently of this direct injury, the wind is no less formidable for another reason, namely, on account of the dust which is then raised from the ground and maintained in the atmosphere breathed. No great harm will be done by the frequent occurrence of a moderately strong wind in a region which is free from dust, but in places with a dusty soil this inconvenience is almost prohibitory, and diminishes to a notable extent the medical value of a station. Certainly it is often possible to diminish this unfavourable condition, by judiciously using for the position of residences such resources as the region may offer. This, however, is but an uncertain course to adopt, and places in which the wind blows but infrequently, or at least without dust, must always be considered as undoubtedly preferable.

The usual hygrometric condition of the place should be taken into serious consideration, not only with regard to its uniformity, but also to its mean degree. The importance of the first condition has already been mentioned, and is easily understood, since any sudden and pronounced variation in the meteorological phenomena is an obstacle to local preservation, and liable to cause its suspension. As regards the hygrometric condition, the degree should notably exceed the middle point of a centesimal measure. At 55, 60, or 65,[*] a climate would be quite or almost dry; it is between 70 and 80 that the desirable limit of mean relative humidity must be sought, the first number not reaching while the second exceeds the desired amount. The effect of dryness of the atmosphere is completely different, according as it is associated with a low temperature, or one which is temperate or hot, and hence these conclusions are opposed to those which were made in reference to the highest climates of the first group. In the mild climates of the valley the usual low condition of the hygrometric measure, or dryness of the air, favours and promotes irritation of the respiratory mucous membrane. On account of the abundance and rapidity of the cutaneous and pulmonary evaporation, a notable quantity of water is removed from the organism, and in consequence the normal and pathological liquids existing therein are concentrated. The truth of

[*] The quantity of aqueous vapour which the air would contain if saturated at a given temperature being represented by 100.

this fact is shown by the expectorated products, the viscous character of which renders expectoration difficult, and by the secretion of the kidney, which becomes of small amount, concentrated, and of high colour. Vivenot, who has specially analyzed these effects, believes that this diminution in the secretion of the kidney may cause, on account of the small amount of water thus removed, the retention in the blood of a certain quantity of urea, which is then eliminated by the skin and lungs in the form of nitrogen, or carbonate of ammonia, such elimination being a direct cause of irritation in those organs. Nor is this all; notable variations of temperature invariably coexist with dryness of the air in climates of the plain, while a somewhat moist atmosphere would be accompanied by uniform temperature. Thus it may be said that a low hygrometric number is not only unfavourable from the bad effect which the atmosphere will then have directly upon the organism, but also from its being constantly associated with much variation in the height of the thermometer. These direct and indirect effects produced by dryness of the atmosphere are all more pronounced in proportion as the temperature is higher, and consequently climates which undoubtedly present these characters in a high degree should only be recommended when indications of a special character exist.

The effect of the atmosphere upon the organism should also be taken into consideration, or rather the impression which it produces upon the mucous membrane of the respiratory tubes. Thus no effect of any kind may be perceived, or a soft and agreeable impression may be recognized, as air enters the chest, or perhaps a disagreeable sensation of roughness is equally well appreciated. This difference is often combined with notable change in the climatic conditions, specially as regards the hygrometric condition and vicinity of the sea. The same distinction, however, may exist between two places in which the closest analogy exists in every other respect.

The special quality of the atmosphere can therefore be its only cause, and this property, which at an extreme degree produces the so-called soft or rough air, cannot possibly be neglected in appreciating the relative conditions of local preservation.

Thus uniformity of temperature, as shown by the slight extent of variation which occurs during the day, from day to day, or month to month; uniformity of the hygrometric condition within the limits of mean humidity; rarity of winds, and absence of dust; softness, or at least absence of roughness in the air, are the climatic conditions necessary to produce the most beneficial effects as regards subsidence of local irritation, and preservation from inflammatory complications.

The second effect of these climates will now be considered, namely, repair of the constitutional condition.

Analytical dissociation, which enables such a fundamental question as improvement of the patient's general condition to be separately considered, also supplies the clearest answer from a theoretical and general point of view. With respect to this, the most important climatic condition is the mean temperature of the place inhabited; the lower the temperature, the more certain and rapid is the improvement which occurs; and if purity of the air and absence of wind are combined with a low temperature, causing the climate to resemble that at a high altitude or on the mountains, such conditions are the best which can be desired in regard to the constitutional condition, and, in my opinion, no doubt can be felt as to their therapeutic effect, when considered separately and in an abstract manner.

In practice, however, the answer is necessarily less decided. Patients certainly exist in whom the constitutional indication is so important as to take the first place and deserve the only consideration. In this case the above theoretical becomes identical with the practical solution of the difficulty, and the most appropriate treatment can be applied without hesitation.

Usually, however, the course to be pursued is less evident, the physician not being justified in concentrating his efforts to improve the general condition, since the local disorder requires an equal degree of attention. The necessity of observing both conditions not only leads to the coldest climates being utilized, as indicated by the constitutional condition, but also those of which the atmosphere is soft, and which therefore answer more completely to the twofold requirement of the indications. It is thus absolutely necessary, this being perhaps the most important question, to understand, as regards the whole group of stations in the plain, and not only some of the individual places in it, what is the real effect of the climate, both as regards strength and constitutional repair.

There are three ways of ascertaining this point. Firstly, the natives of the place, and those who habitually reside there, may be consulted. This, however, is the worst plan to adopt, on account of some errors of appreciation which may result from that connection, whether by birth or residence. The condition and knowledge of these places derived from patients who have resided therein for some months, as well as from persons in good health who accompanied them, afford a second means of learning these facts—a plan which is much better with regard to the value of the knowledge gained, but most defective in its application, the patient having been exposed to all the dangers of a trial from which he should have been protected by the previous experience of his medical adviser. Lastly, the plan of obtaining a personal impression of the place remains, which is far preferable, and unites certitude with rapidity. But a short time is needed by one who has the power of making observations, that the strengthening or debilitating effect of a climate may be recognized, so that the conclusions made are directly applicable to the patients. They may even be applied with a force increased by the whole difference which exists between a healthy and diseased organism in its power to resist outward impressions. The effect upon a person in good health considered alone will always be similar to that produced upon a patient affected by phthisis, but will be greater in the latter case on account of the weakness and susceptibility of the person. It is to the appreciation of these facts and to the formation of this opinion that I alluded at the commencement of this work, when speaking of the special effect of climate upon the organism under its influence, the indefinable action of which cannot be explained in all cases, and would be vainly searched in the table of atmospheric conditions. This action, it has been already said, can only be appreciated in the place itself, and must be felt to be understood. Though the information obtained from patients and others who resided with them is by no means without value, this should be preceded by a knowledge of the place, and such association of personal opinion with data resulting from pathological observation constitutes, in my opinion, the best means of acquiring knowledge of the place. It is upon this twofold basis that the practical opinion and advice are founded which will soon form the subjects of discussion.

If the importance which has been ascribed to constitutional malnutrition as regards the origin of the disease be borne in mind, and the great value which should consequently be attached to the general condition of the patient, it will easily be understood that the question whether the effect is to strengthen or debilitate should hold a large place in the consideration of climatic treatment, and that this element of judgment should be the subject of constant thought.

Before more is said, however, a mistake should be mentioned which is often made, and which should be avoided.

In the question of climatic treatment the exciting or strengthening effect of a climate on the one hand, and the weakening or sedative effect on the other, are often confused, and to such an extent that in both cases the designations are indifferently used one for the other. This, in my opinion, is more than confusion, being a real error. The exciting or sedative effect is dependent upon the condition of the nervous and vascular system, and upon the presence or absence of irritation in the respiratory tube, while the fortifying and tonic, or debilitating effect, depend upon the nutrition and degree of strength in the organism of the patient.

The accuracy and necessity of this distinction may be shown in the following way. What is the result of the above-mentioned effects upon the patient? Excitement shows itself in physical and moral agitation, in susceptibility which is so pronounced as to be painful, in continuous insomnia, often in headache, acceleration or irregularity of the respiratory and circulatory rhythm, and lastly and undoubtedly in irritability of the larynx and bronchial tubes. With this deceptive appearance of movement and life are associated sooner or later a sensation of real and constant weariness, an increasing aversion to all exercise which is combined with fatigue, and often more or less pronounced diminution of nutritive activity. When the effect of the climate is fortifying none of these changes occur, the functions of the nervous and circulatory system remaining as before, the irritability of the respiratory mucous membrane being in no way increased, while the patient experiences a feeling of energy and of new force. This sensation is due to some real change as shown by the daily increasing aptness for muscular exertion; the increased appetite, improved digestion and assimilation, the activity of the hæmatosis, also characterizing this condition, which is proved to exist by the increased strength and weight of the body. Two more opposite conditions, two groups of effects more different from each other, cannot be conceived, and a similar distinction exists between sedative and debilitating action. The confusion mentioned, which is opposed to the true condition of the organism, should, therefore, be avoided, and two modes of action, of which the most pronounced types have already been stated, should be carefully distinguished.

The two properties which might be termed positive, that is to say, the exciting and the strengthening effect, may be united in the same climate, and the negative properties, viz. the sedative and the debilitating action, may occur in the same region. Such an association, however, is by no means frequent, and consequently their being looked upon as similar or synonymous terms causes the action of many climates to be erroneously interpreted. Thus, for example, Egypt,* which is, undoubtedly, the most exciting climate of the second group, is also the most debilitating, as I understand the word. Madeira,† again, of which the climate is specially sedative, far from inducing debility, has, on the other hand, a fortifying effect, and I am glad to find that my eminent colleague, Lombard (of Geneva), agrees with me upon this point. Other instructive examples may be given. In Algiers,‡ or rather Mustafa, it will be found that the two properties are modified in a different sense, the tonic action being greatly diminished, while the exciting action is increased. The reverse also occurs in the climates of the Riviera, upon the shore of the Mediterranean Sea; all these are exciting in different degrees, and though the synonymous use of the terms would represent them as having a strengthening effect, this, in my opinion, is not the case, and I should regard them as indifferent, and as having no effect upon the strength of the patient. The Riviera,§ on the Italian side, has literally no invigorating effect, while on the French side this, if at all the case, is so little pronounced that it cannot be considered an element of real therapeutic importance. Such, at any rate, is the definite conclusion which I have formed upon this subject.

Sometimes, as has been already observed, the analogous

* See note, p. 297.
† See note, p. 243.
‡ See note, p. 297.
§ See note, p. 324.

properties exist together. Thus, for example, it is certain that the climates of Méran * and Montreux † are not only strengthening to a high degree, but also exciting. Those of Pau ‡ and Pisa,§ again, are remarkable both for their sedative and debilitating effect. An attempt has, I know, been made latterly to show that the climate of Pau has a tonic effect, but experience prevents me from accepting such a conclusion. Thus there are certainly climates with analogous properties which act in a similar manner, and with more or less effect, so to speak, upon the nutrition and strength of the patient. Even then, however, my distinction should be maintained, since it is very rare that these analogous properties are equal in degree. This is specially the fact as regards the stations of Méran and Montreux, the exciting action being so slightly pronounced in comparison with their tonic effect that it may be entirely disregarded, the constitutional result being in reality the most important consideration.

The distinction which has just been dwelt upon is, therefore, fundamental; if neglected, not only will the effect of climates be misunderstood, but the patients will be ill directed. What, in fact, is the inevitable consequence of this mistaken identity between an exciting and a fortifying effect? Is it not the equally erroneous fusion of two therapeutic indications which differ completely from each other, namely, that furnished on the one hand by the excitability of the nervous and vascular system, and on the other by the good or bad condition of strength and nutrition. The first indication is purely symptomatic, while the second is connected with the cause and pathology of a disease which originates in and is maintained by constitutional malnutrition. This simple statement shows that the clinical value of the distinction is at least equal to its interest as far as the climate is concerned, and that of

* See note, p. 320.   † See note, p. 325.
‡ See note, p. 309.   § See note, p. 320.

these, one is incomparably more important than the other. There are several reasons why the preponderating importance of the indication furnished by constitutional debility should be well understood: firstly, from its being connected with the cause of the disease; secondly, from its being constant on that account; and lastly, from its effect being uniform and always requiring the same plan of treatment. The indication drawn from excitability in the patient is, on the other hand, of secondary importance, being a symptomatic indication of inconstant existence and of varying practical signification.

It results from this, that of the two climatic properties which I distinguish, the fortifying action is by far the most important, answering to one of the fundamental indications of the disease, as an agent of local protection. As was said at the commencement of this work, the only solid basis of prophylactic and curative treatment is furnished by the notion of malnutrition, and by knowledge of the injurious effects of inflammation. What is true of the general is no less so of the climatic treatment, which should always and chiefly be influenced by two causal indications which exist in every case, but of which the respective importance and urgency differ in each patient. The constant and immutable conclusion is, that though the symptomatic knowledge furnished by the excitability of the patient may be so pronounced as to deserve consideration and treatment, the two principal indications must still be as far as possible followed.

Local protection and constitutional repair are thus the two possible effects of climates in the second group. Meteorological uniformity in every sense, with absence of dust as respects the local, and fortifying action with little or no exciting influence as regards the general effect—such are the conditions necessary that the desired result may be obtained.

The best climates, then, are evidently those which satisfactorily answer to these two therapeutic indications. With this exception the most beneficial are those which, though less well adapted to fulfil these two requirements, are specially appropriate and efficient as regards one of them; the climates which have been termed indifferent, on account of their characters being less accentuated, or owing to some special condition, do not respond with certainty or advantage to either indication. Without being opposed thereto, they are but vaguely applicable, and their therapeutic value on this account alone is inferior to that of the two first groups.

The practical consequence of this division will have been foreseen. The climates in the first group, unless other sources of judgment exist, are appropriate to every patient, fulfilling the two indications which exist in all cases; the stations of the second group are adapted to those in whom one of these fundamental indications is undoubtedly more urgent than the other, meteorological uniformity, perhaps, or tonic action being specially required; those of the third group, from their undecided effects, are adapted to patients of both kinds, but from their slight influence should only be utilized when nothing better can be done.

The new division of climates which I have made, and which is founded upon the varying relation between the action of the climate and the two ætiological indications of the disease, is, in my opinion, the most useful as well as the most medical classification. It is certainly rare that these two indications need be alone fulfilled; in many cases others exist which should be considered, such as excitability, the mode of reaction, or, again, some constitutional condition, such as rheumatism or scrofula. These occasional indications, however, should not be followed to the prejudice of the two others, which are in every case of preponderating importance. In these cases of complex and at times irreconcilable requirements, climates must often be renounced which are perfectly adapted to the fundamental indications, and some plan be adopted which is appropriate to the different necessities of the case. This arrangement, however, should never be accompanied by neglect or violation of the two primary indications. Thus it is both legitimate and logical, from a medical point of view, to take as the basis of division with respect to climates of the plain their respective action as regards local protection and constitutional repair.

After this analysis of the necessary climatic conditions, it would be easy to illustrate by three series of names the truths which have just been established. This, however, will not be done for two reasons: firstly, because the dogmatic character of this proceeding renders it inexpedient; secondly, because such an enumeration would not altogether elucidate the subject, leaving all secondary indications unconsidered. It will, therefore, merely be said that the types of my first series,* with some shades of distinction, are Madeira† and Algiers;‡ while in the second, Méran,§ Montreux,‖ and Lugano¶ are types of the climates which are specially appropriate on account of their invigorating action; Palermo** and Catania†† having rather a protective effect as regards the local affection. Lastly, Pisa‡‡ and Pau§§ should be mentioned as types of the third series. This must be regarded as a mere statement in order to fix the ideas.

It should be mentioned that Madeira and Algiers, the types of the first series, should also appear in the second, on account of their double effect. Since, however, the power which they have as regards local protection is greater than that of their tonic influence, these two stations will hold a different place in the second series, according to how they are regarded. Their constitutional action, though similar, is far

---

* These three groups of climates are those mentioned on pp. 367, 368.
† See note, p. 243.   ‡ See note, p. 297.   § See note, p. 320.
‖ See note, p. 325.   ¶ See note, p. 320.   ** See note, p. 328.
†† Ibid.   ‡‡ See note, p. 320.   §§ See note, p. 309.

inferior to that of Méran, Montreux, and Lugano, whilst they are far preferable to the stations of Sicily as regards local protection.

Having concluded this introduction, which was necessary that my personal views as to the action of climates in the plain might be explained, the pathological groups will now be considered, and the relation which they bear, in my opinion, to those which are climatic. The same difficulty, it will be understood, exists here as was encountered when the high climates and mineral waters were being mentioned. A dogmatic statement might, perhaps, rigidly speaking, fix the corresponding climatic and pathological groups, but the selection of an individual station in the group chosen is so variable as to defy, in most cases, all general rules. As complete information, however, as the subject permits will now be given.

Before commencing, it should be once for all stated whence the meteorological numbers stated in this chapter are derived. Neither borrowed, nor copied from another work, they are all the result of personal labour, in which I utilized, firstly, information obtained at the place itself, and secondly, French or foreign publications of a scientific character. The study and comparative appreciation of these documents, and the information gained by me, are the origin of the numbers which are here produced, and which are, therefore, my own, and should be accompanied by the name of their author. Being the expression of a great number of different elements, they differ from those which are contained in any isolated work, but their exactness can be guaranteed from the character of the various sources whence they are derived.

In all the meteorological data, unless the contrary is stated, the Fahrenheit (Centigrade) scale is used, and they belong uniformly to the period which extends from October to April. This uniformity, so often neglected, is an indispensable condition that the comparison may be exact and profitable.

In the first place, a pathological group will be considered which is clearly defined, and with respect to which the climatic equivalent may be found with a relative facility which is quite exceptional. Allusion is now made to the so-called "pneumonic phthisis" when it has reached such a chronic stage as to be compatible with the patient's removal from home.

With the one rare exception mentioned when the treatment by high altitude was being considered, this form of the disease always indicates the same climatic group—a group which includes but four stations, namely, Madeira,* Algiers,† Palermo,‡ and Pisa.§ Pau ‖ could scarcely, in my opinion, be joined with them on account of the larger extent of variation in the temperature, not only during the day, but from day to day, and from month to month. What is of the greatest importance both as regards freedom from danger and the benefit of treatment is, above all, meteorological uniformity; if combined with action which is tonic rather than debilitating, all is at its best, and the desirable ideal is realized. This fortunate combination is found in different degrees at Madeira and Algiers— not existing at Pisa, which is only favourable on account of its uniform climatic conditions, this uniformity, however, being notably less than at the two other stations.

Whenever the physician is at liberty to select any station in the pathological group, which, unfortunately, is by no means usual, the preference should, in my opinion, be given to Madeira, for the following reasons. With a mean temperature of 64·2° F. (17·88° C.) for the seven months between October and April, a mean variation of 6° F. (3·33° C.) from day to day, so trifling as from 1·1° F. to 1·2° F. (0·65° à 0·70° C.) during the day, and a mean monthly variation of 1·5° F. (0·83° C.), or at the most 4° F. (2·2° C.), Madeira realizes more than any

* See note, p. 243.   † See note, p. 297.   ‡ See note, p. 328.
§ See note, p. 320.   ‖ See note, p. 309.

place in our hemisphere, and without any comparison, the different conditions requisite both as regards climatic uniformity and the degree of temperature which exists.

In addition to this unequalled advantage which it is illogical to neglect, there are others which, to my great surprise, have never been stated, notwithstanding the serious consideration which they undoubtedly deserve. The rule enjoining prolonged residence, which is never more imperious than in pneumonic phthisis, can be observed without the least difficulty, owing to the slight difference between the mean temperature in summer and winter, viz. from 9° to 11·5° F. (5° à 6·45° C.), and owing to the number of inhabitants who live in the vicinity of Funchal from the sea-level to the height of 2150 feet (650 *mètres*) or more. Conditions, in fact, exist which have no analogy elsewhere, and which enable the numerous and dissimilar indications to be followed which result from individual excitability and the mode of reaction. The same circumstances prevent the climate of Madeira from being regarded as uniformly sedative in effect, which would not be done if its nature was truly understood. At the level of the sea, or almost so, it is certainly sedative without having any debilitating effect. When, however, a greater height is reached, the sedative can be replaced at will by undecided, and then by slightly exciting action, and this is combined with a pronounced tonic influence due on the one hand to the height of the place, on the other to the low temperature, which falls in the mean 1° F. for every ninety feet of elevation (*un degré C. par cinquante mètres d'élévation*).

Nor are these the only valuable advantages of this station. The town of Funchal * and the fine amphitheatre of villas which surround it, to a height of almost 2300 feet (*près de* 700 *mètres*), occupy the southern aspect of the island, and are completely

---

* Funchal is the capital town of Madeira, on its southern coast. See note, p. 243.

protected from the northern, north-easterly, and north-western winds by many mountain ranges, which reach, and even exceed in more than one place, the height of 6560 feet (*deux mille mètres*). The whole place is supplied with potable water, remarkable both on account of its purity and abundance; and, lastly—an advantage which, in my opinion, is by no means inappreciable—dust is completely absent both at Funchal itself and in its vicinity. This statement is the precise truth. Though the whole day may have been passed in the open air, there is not a particle of dust upon the clothes. The soil is volcanic, and the roads, whether of large or small size, are paved with small stones closely pressed together for some distance beyond the inhabited zone, so as to resemble basalt in mosaic. Even when the wind from the sea reaches its greatest intensity no dust can be raised, and the atmosphere preserves its unalterable purity. The excellence of the hotels and villas as regards comfort, and other requirements already mentioned, need not be dwelt upon, and the not unimportant fact should be also noted, that mosquitoes are entirely absent.

It is unnecessary to mention the indescribable charm of this mountainous country, a true Switzerland in the ocean, and the enchantment produced by a luxuriant vegetation, which presents the riches of a tropical flora combined by simple association with products of the temperate regions; this is due to the fact that in this work the useful is alone considered, without reference being made to the agreeable, though in this disease the latter often constitutes an element of the former.

If, notwithstanding the easy journey, the patient refuses to live at Madeira on account of its distance, Algiers [*] and Mustafa [†] should without doubt be the places recommended. In making this change, the advantage of a fixed residence will be lost, as also the valuable effect obtained by variation of altitude, while the advantages will be obtained of a climate

[*] See note, p. 297.   [†] Ibid.

whose action is not too exciting, which is fortifying rather than the reverse, and of which the uniform temperature, though less exceptional than that of Madeira, is still of a remarkable nature. The mean daily oscillations of the temperature, during the seven months from October to April, do not exceed 8·1° F. (4·5° C.), the mean monthly variation is of 3·5° F. (1·93° C.), and the mean temperature 62° F. (16·62° C.) during the same period. That all these favourable effects may be completely obtained, the residence of the patient must be in a place which is completely sheltered, as at more than one spot about half-way up the ascent of Mustafa Supérieur.

In addition to this, it would not be fruitless to rectify an error which is but too far spread respecting the exposed position of Algiers and Mustafa, a too superficial examination having given rise to the statement that this town looks to the north. This is incorrect, and arises from the fact that the large contour described by the western part of the bay has not been remembered. The aspect of Algiers is in reality directly eastward, and that of Mustafa is to the east-south-east on the western side of the slope, while on the central and eastern side it is to the north and north-west. On the western side the shelter from the winds is as complete as could be wished, owing on the one hand to the presence of the mountain of Bou-Zarea and the Emperor's fort (Fort de l'Empéreur), on the other of the chain of the Atlas and of Djurdjura. As these remarks would indicate, it may be truly said that uniformity of the temperature and purity of the atmosphere are realized at Mustafa in such a satisfactory manner that it is inferior to no place but Madeira in these two respects. It should be added that in these two regions, and in them alone, there is complete protection from the inconvenience of dust. The character of the climate may be completed by stating that it is more exciting than that of Madeira, having at the same time a somewhat less strengthening effect. It is, in fact, the least tonic of all the climates which in my opinion deserve that qualification, the others being those of Corfu,* Catania,† Palermo,‡ and Madeira,§ this enumeration being made in order of decrease. Thus the station of Mustafa is seen to be most appropriate to the special indications furnished by the pathological group which is now being considered, and while a slight difference in the excitement and fortifying effect produced distinguishes it from Madeira, the most important varieties of excitability and individual debility may be satisfied at both places. It is still to be regretted that even at the present time Algiers cannot well be inhabited during the whole year, the change of altitude not being accompanied by the necessary alteration in the comforts of life.

When pneumonic phthisis, besides becoming chronic, which is the *sine quâ non* of climatic treatment, is undoubtedly of an indolent type, this combination, which is but rare, definitely indicates residence at Palermo. This station is completely protected from the northern winds, while the suburbs of the town, and notably Conca d'Oro, furnish valuable opportunities of residing in country air. The climate, as I regard it, is in some degree more exciting and more invigorating than that of Madeira and Algiers, having the same uniformity as in the latter region, specially from the fact that the daily oscillations of temperature do not reach 9° F. (5° C.), which, in my opinion, should be the limiting amount. The mean temperature is notably lower, being 56·8° F. (13·79° C.) for the seven months from October to April. The town and its immediate neighbourhood are not liable to the inconvenience produced by dust or wind; but, as with regard to Algiers,

---

* Corfu is an island in the Mediterranean, at the entrance of the Adriatic Sea, and near the coast of Albania. See "Handbook for Travellers in the Mediterranean Sea" (J. Murray); "Handbook for Travellers in Greece and the Ionian Islands" (J. Murray); "Influence of Climate on Pulmonary Consumption," C. J. Williams; "European Guide-book," Appleton; etc.

† See note, p. 328.   ‡ Ibid.   § See note, p. 243.

it is quite impossible to reside there during the winter months. The beauty of the position and bay of Palermo is proverbial, and need not be dwelt upon.

Lastly, if the pneumonic phthisis, whilst being chronic and inactive so far as the local foci are concerned, coexists with an individual condition of the same torpid character, it is Madeira which should be selected, to the exclusion of Algiers and Palermo; or, if an insuperable repugnance exists to such a distant place, Pisa,* which, with the exception of Madeira, is specially appropriate to these several indications. Substitution of one place for the other, however, does not mean that their qualities are similar. Whether the daily oscillations, the changes from day to day, or from month to month, are considered, the uniformity, though more decided than at Pau,† is less so than at Madeira, the mean daily oscillation reaching the limiting amount of 9° F. (5° C.), whilst at Madeira it is only 6° F. (3·33° C.), the mean temperature being also lower, namely, 50° F. (10° C.) during the seven months from October to April. Thus the conditions are altogether inferior, for which reason Pisa has been ranged among the indifferent climates. In the supposed case, however, it does not seem that a better choice could be made. Notable improvements have also been made at Pisa during the last few years, specially at the hotels, the houses occupied by the patients, and in the medical resources of the town, which contains also at the present time a most complete establishment for hydro- and aërotherapeutics.

It is not extremely rare that patients affected by pneumonic phthisis have suffered, or perhaps are still suffering, from more or less pronounced rheumatic symptoms. This circumstance notably limits the choice of residences, since in such a case, whatever indications may otherwise exist owing to the excitability or individual reaction of the patient, Palermo and Pisa must be absolutely rejected, though their hygrometric

---

* See note, p. 320. † See note, p. 309.

mean is far superior to that of Madeira and Algiers. The two latter stations, on the contrary, are notably analogous in this respect, whether the mean hygrometric condition during the seven months from October to April be taken into consideration, the number of days upon which there is rain, or the quantity which falls during the same period of time.

For some years, especially in England, the island of Teneriffe,* in the Canary Islands, has been highly extolled as a residence for patients affected by phthisis. This island, situated to the south-east of Funchal,† can be reached from it by a sea voyage of twenty-eight hours, and as the climate is more dry and exciting, residence in that place will complete and extend the therapeutic applications of Madeira. In numerous cases, in fact, it should be preferred, specially when patients are affected by rheumatism, or in whom the reaction is but slightly pronounced. If the climate be alone considered, nothing is more true than these statements; the island of Teneriffe, or, more precisely, the valley of Orotava in that island, fulfils the greater number of the necessary requirements to the same extent as Madeira. The mean heat of Orotava during the period of time between October and April is even higher, reaching the temperature of 65·5° F. (18·52° C.). There is the same uniformity with respect to the daily and monthly oscillations of temperature, except that the daily oscillations are slightly more pronounced, the difference being so slight that it may be neglected. It may be added that the configuration of the island offers the same facilities with respect to prolonged residence and variations of altitude, the latter reaching a high elevation, so as to produce all the effects of a mountainous

---

* Teneriffe is the largest and most important of the Canary Islands. See "Principal Southern and Swiss Health Resorts," W. Marcet, M.D.; "Les Iles Canaries et la Vallée d'Orotava," etc., Gabriel de Belastel; "Voyage aux Régions Équinoxiales du Nouveau Continent," vol. i., A. von Humboldt, and its translation (Bohn's Scientific Library).

† See note, p. 372.

climate, with the same rarefaction of the air; conditions being thus united together which are specially favourable in a station with such varied applications.

Knowing these particulars which have such a distinctive character, and aware of the favourable opinion of our English colleagues, I visited this island myself for the purpose of appreciating its value as a medical station. In my opinion, it is impossible to assign this character to it. The conditions of the climate are as above stated, and the beauty of the valley of Orotava is remarkable. Humboldt (from whom, however, I differ—a fact of little importance) states that this is the most beautiful place in the world.* The dust and mosquitoes, however, are in permanent possession of the country, in addition to which there is not in any part of the island, either at Orotava or elsewhere, a single establishment which is suitable to patients; the food supplied is of bad quality, and no improvement whatever can be expected, since the inhabitants decline, almost with indignation, to adapt their country to the requirements of consumptive patients, and will on no account agree to such a transformation. The question is, therefore, settled, and the island of Teneriffe must be removed from the list of medical stations, notwithstanding the exceptional advantages of its climate, so long as the dust and mosquitoes constitute the objection which they do at the present time.

The absence of suitable establishments, and the imperfection in the public hygienic arrangements, and in those connected with food, notwithstanding statements made in the opposite sense by English physicians, prevent me from regarding Mogador and Tangier,† to the west and north of

* Humboldt's words as given in the narrative of his travels are, "J'avoue—n'avoir vu nullepart un tableau plus varié, plus attrayant, plus harmonieux par la distribution des masses de verdure et de rochers" ("Voyage aux Régions Équinoxiales du Nouveau Continent," vol i.).

† Mogador and Tangier are towns in Africa, upon the coast of Morocco. See "Health Resorts," J. Burney Yeo, M.D.; "Handbook for Travellers to the

Morocco, as medical stations. There also, and specially at Mogador, the climate, as in the island of Madeira, is most appropriate; but this is naturally useless if no other advantage exists. Such is the case at present; but Mogador differs from Teneriffe in the fact that the required improvements may in time be made, and remove the reasons which now exist for condemning it. In fact, during the last year, I myself saw constructions at Tangier in an unfinished condition, which, and specially on account of their position outside the town, indicate that real progress is being made.

At the present time, however, Teneriffe, Mogador, and Tangier are only beneficial on account of their climate, and it has already been fully shown that this alone will not suffice to establish a medical station.

Phthisis in the ordinary form will be now considered.

The precept which, in my opinion, should be adopted is as follows:—As long as possible, consumptive patients should be enabled to derive benefit from the healthful effects of invigorating climates. Such is the case whenever the constitutional is more serious and urgent than the local indication, as is most frequently the case. Whether there exists a catarrhal affection limited to the apices, or circumscribed and more indurated foci, formed without the occurrence of any febrile attack, is of little consequence. The indication is in all cases the same. The mild climates should not, in my opinion, be employed unless some accidental modification occurs in the picture which has just been presented, since, as already stated some years ago in my Clinical Lectures, it is quite illogical to abandon, with respect to the climate occupied, the fundamental indication which is to discipline and invigorate the constitution.

Now, in the group which contains climates of the plain,

Mediterranean Sea" (J. Murray); "Influence of Climate on Pulmonary Phthisis," C. J. Williams; etc.

are three stations, separated by their mean temperature from those with mild winter climates. Without having the rigorous weather of extreme altitude, this is certainly not mild in character, and they thus form, in the series of places in the valley, a distinct group which may be considered intermediate. Such are Méran,* Montreux,† and Lugano,‡ of which the respective mean temperatures are comprised between 42·8° F. and 44·6° F. (*entre* 6° *et* 7° C.) during the seven months from October to April. On account of the invigorating action which these climates undoubtedly have during this period, their effects are analogous to the climates which are termed rigorous, but without the special effects of altitude, which in their case does not exceed 1260 feet (385 *mètres*). Thus, while less invigorating than those of the first group, they have without comparison the most tonic effect of all climates in the plain. On the other hand, they produce no excitement, or, to speak more exactly, their exciting effect is too slight, specially at Méran, to cause any serious anxiety, or form any fundamental element in the selection of a residence.

It would seem, therefore, at first that these three stations possess a preponderating and exclusive adaptation to the pathological group under consideration. Such, however, according to my experience, is by no means the case, as it will not be difficult to show. But for the absence of altitude these stations should be ranged in the first group, since on account of their general climatic characters in winter they much resemble the places in that series; hence it is that they are not really appropriate to any special form of the disease, fulfilling in reality the same indications as stations which are highly situated. Since the latter have in the advantage of high altitude, combined with the possibility of carrying on methodical treatment, and all the hygienic and medical advantages which distinguish the sanatorium of the hotel, they

* See note, p. 320.   † See note, p. 325.   ‡ See note, p. 320.

ought certainly to be preferred. It has been already said, but may well be repeated, that consumptive patients who are able to live in the climate of Méran, Montreux, or Lugano can also tolerate Davos,* the Engadine, or at any rate the lower stations of the first group; in my opinion, therefore, it is to these places that they should be directed, Méran, Montreux, and Lugano being reserved as supplementary stations for patients who are absolutely prevented by personal reasons from residing in places which are more highly situated and have more characteristic effects.

It should in justice be said, that these places in Switzerland and the Tyrol † present in the same way as Davos and the Engadine every desirable facility for employing the methodical plan of treatment during a prolonged residence, and offer appropriate abodes in their immediate vicinity, at heights which vary from 1970 to 3280 or 3940 feet (*de* 600 *à* 1000 *ou* 1200 *mètres*). Hence it is that at the end of spring the patients can escape from the summer temperature while still remaining in the same climatic zone, the altitude being alone changed. One of the elements which causes the superiority of Madeira is found here in quite a different sphere, and it would be vain to seek for another equally complete and characteristic example.

Although very analogous, the stations of Méran, Montreux, and Lugano are not so similar to each other that they may be chosen with indifference. Considerations furnished either by the patient or station will constitute in each special case reasons for preferring one or the other; but these are questions which regard the individual climate as well as the morbid tendencies of the patient, into the details of which it is impossible to enter. At the same time, it should be said that,

---

* See note, p. 243.

† The Tyrol is a country forming part of the Austrian dominions, to the east of Switzerland and south of Bavaria.

with respect to the hydrotherapeutic and pneumatic treatment, Méran contains excellent establishments which have no existence in the Swiss stations. The three places, on the other hand, are equally to be recommended when the grape-cure * is being employed, which in some cases is of real benefit in this disease.

When for any reason the onset of the complaint has not been utilized for the purpose of rendering the climatic treatment, which has such an invigorating and salutary effect, available to the patient; when the affection is at an advanced stage, the lesions being more considerable, and serious in themselves; or, again, when, since the first development of the morbid growth, the patient has presented an obvious tendency to be affected by congestive and inflammatory attacks leading to acute and febrile complications, the condition of things is no less clear, and the time has come for employing decidedly, and probably without cessation, the softer climates and southern stations. Protection from local complications is no less urgent than constitutional restoration.

The fundamental principle according to which selection should be made between the different places, is the mode in which the reaction occurs, to which, in my opinion, the greater or less tendency to febrile complications should be ascribed. This consideration, in my opinion, overrules everything else, even the occurrence of hæmoptysis, since, as has been already shown, this is of variable signification with respect to the prognosis, and requires such treatment as is specially appropriate to the form of individual reaction.

Without doubt—and this must be clearly understood—attention should be also paid to any laryngeal or intestinal complication of serious character, or to any constitutional

---

* The grape-cure is the systematic employment of grapes as a means of cure. See "The Curative Effects of Baths and Waters," Julius Braun; "Baths and Wells of Europe," J. Macpherson, M.D.; etc.

condition which may be specially apt to be associated with phthisis, scrofula, or rheumatism; this, however, is of secondary consequence, and though important, should on no account take the place of the preceding indications. Their effect may be to settle the individual climate in the group of stations indicated by the mode of reaction, but cannot and ought not to change the group selected. Consequently, as regards those cases which are now considered, and which, unfortunately, are but too numerous, the primary indications of constitutional restoration and local protection being equally and constantly in existence, a classification of the climates which is really practical can only be based on their being more or less truly appropriate to the different modes of individual reaction.

Of these there are, in my opinion, three, as was explained at some length in the earlier part of this work. The distinctive terms which were applied to them need be alone recalled, namely, those of active or florid, passive or torpid, and undecided or ordinary reaction. Such are the three modes of reaction which may occur in any patient affected by phthisis. Hence three climatic groups are clearly specified, at any rate as regards the two first forms. Whichever group is indicated by the individual character of the patient, a place should be chosen in it of which the special character is most decided. So long as the group is well constituted and selected, such climates must certainly be more effective in attaining the double object than those which, while belonging to the same group, have less pronounced effects. This rule of action is, as far as I am concerned, a general and absolute law, from which I never deviate unless the patient refuses to abide by it. As no benefit can be then produced, the least harm possible should be done, though the fact is to be regretted, since experience shows the patient to suffer loss by such inopportune treatment.

These, then, are the rules of my practice; their application will now be taken into consideration.

When the reaction is active, the indication is in all respects similar, so far as climatic treatment is concerned, to that of pneumonic phthisis; the indications furnished by the nervo-vascular excitability, the liability of the bronchial tubes to disease, the previous existence of more or less frequent and acute complications, are identical in the two conditions, and the same group of stations should be therefore utilized in the two cases. It is thus Madeira,* Algiers,† Palermo,‡ or Pisa § which should be recommended, the first of these stations being adapted to all kinds of excitability, while Palermo and Algiers should be exclusively reserved for those forms of disease in which the reaction is but little pronounced. This distinction being recognized, the selection between the three latter stations becomes an individual question to which no definite answer can be given. In a general way Algiers and Palermo should be preferred, in my opinion, on account of the principle already laid down with regard to climates which have the most decided characters; when, however, the patient is affected by some decided laryngeal complication, or has recently suffered from an acute affection, I recommend Pisa, should the patient be unable to reside at Madeira. With respect to the constant superiority of the latter station, this results, as already stated, not only from its climatic conditions and special adaptation to long residence, but also from its appropriateness to the different forms of reaction by means of variation in the altitude. No station could be conceived which more fully answers to all the indications; but if it is wished to derive from it every advantage possible, the residence there should be prolonged, and from eighteen months to two years is, in my opinion, the most advantageous period of time to reside in this island.

* See note, p. 243. † See note, p. 297. ‡ See note, p 328. § See note, p. 329.

The ordinary form of phthisis, even when the reaction is pronounced, exposes the patient less than the pneumonic form to the development of severe acute complications, which, should they occur, have a very different signification in the two cases. Whilst in the first it is an unfortunate incident, in the second it indicates a return to the acute symptoms of the onset, which may definitely exclude the patient from any benefit which might occur in the chronic stage, and unsettle the plan of treatment already adopted. Hence it is that the necessity of protection from local disease is more certain in the ordinary than in the pneumonic form; and that a perfectly uniform temperature is somewhat less urgent in the first than in the second case. It is for these reasons, in my opinion, that Pau* may be enrolled in the climatic group which is now being considered, though not adapted to the treatment of pneumonic phthisis. The mean temperature of the seven months from October to April is slightly lower than at Pisa, being 47·7° F. (8·7° C.); the variations during the day, from day to day and month to month, are greater, and notably so, specially those occurring during the day, which in the mean are 11·3° F. (6·3° C.), and may reach from 19° to 21° F. (11° à 12° C.). The hygrometric and pluvial conditions, however, are most analogous; and, in short, the climate of Pau, when compared with that of Pisa, has a sedative effect which is less decided, and can be employed when the reaction is pronounced, though not to a high degree. Nor has it the same debilitating action. Hence residence at Pau will be most opportune in patients who, while refusing distant excursions, are affected by such constitutional debility that it may be unadvisable to subject them to the more debilitating effect of residence at Pisa.

It is a mistake which seems frequently made, to associate the idea of pronounced reaction with that of organic strength,

* See note, p. 309.

and from this notion to conclude that cases exist in which the sedative influence of the climate will be sufficient, without it being necessary to take into consideration whether it has an invigorating or weakening effect.

The fact is that in the great majority of cases the reverse happens; the pronounced reaction, the excitability which has but the appearance of energy, is combined with a somewhat decided condition of weakness, and consequently the invigorating, or at least the absence of debilitating, action in the climate is quite as necessary as the sedative effect. It is precisely for this reason that the stations of Madeira, Algiers, and Palermo are of such undoubted superiority in the pathological group which is now being considered, since, while all are sedative in different degrees, they also have a tonic rather than a debilitating effect.

It has been already said that patients affected by phthisis who present, or have presented, symptoms of rheumatism should not be sent to Pisa or Pau. It is unnecessary to dwell upon this point, the rule being absolute whatever the other conditions of the patient may be.

The passive or torpid type of reaction has not for its sole corrective the exciting effect of the climate. The passive nature, the inaction of the organism, is specially due to constitutional malnutrition, and the true and efficient corrective should be sought in tonic action. It is thus, in my opinion, a mistake which would prove prejudicial to patients, to regard the most exciting climates as the best which can be employed when the reaction has a torpid character. It seems that in such cases the principal object should be to fulfil as far as possible the constitutional indication, by selecting climates which have not only an exciting but also an invigorating effect. Catania, and in some cases Corfu, are then, in my opinion, of special utility; the exciting effect is less pronounced at Corfu than at Catania, while the tonic influence preponderates at the former of these two places, the mean temperature of which, during the seven months between October and April, does not exceed 51° F. (10·5° C.), whilst at Catania, during the same period, it is 57° F. (14·12° C.). The oscillations of temperature are also similar except as regards the monthly variations, which are greater at Corfu than at Catania. In both places the mean daily oscillation exceeds the limit of 9° F. (5° C.), reaching almost to 12·6° F. at Catania (7° C.). The respective adaptation of the Greek and Sicilian stations is clearly indicated by these climatic data; laryngeal complications and rheumatism indicating Catania,* since this station is far less damp than that at Corfu. The island of Corfu,† specially the town and its environs, owes the acceptable arrangements which exist there to its temporary submission to English rule. At Catania considerable improvements have been recently made, both at the town and in the surrounding districts. Owing to the careful supervision of the streets and volcanic soil there is no exposure to dust, and at Acireale, situated at no great distance from the town, the same climatic conditions are conjoined with those of a truly rural station. It is useless to dwell upon the marvellous beauty of this region, overlooked as it is by the majestic cone of Mount Etna.

With respect to the patients who are now considered, in whom the reaction is inert and combined with the necessity of as complete protection from local complications as possible, Egypt should, in my opinion, be regarded as the least beneficial place in the appropriate climatic series, a country which is very generally regarded as of the highest value. Without doubt, residence in that country may be most beneficial to patients of an inactive disposition, but is only suitable to a very limited number, and exacts certain precautions which are, as I think, indispensable. The high mean temperature of

* See note, p. 329. † See note, p. 375.

Cairo,* which is 62·4° F. (16·91° C.) for the seven months, the exceptional rareness of rainy days, of which not more than from seven to nine occur during the same period, cause this to be the extreme type of exciting climates, though presenting all the inconveniences of absolute dryness which have been already discussed. With the most exciting action, again, is combined a debilitating effect, the existence of which is, in my opinion, incontestable; and whilst on this account Cairo is directly opposed to constitutional repair, it fulfils but most imperfectly the work of local protection, since, though the mean daily temperature is high, the variations during the day, and from day to day, are so large and abrupt as to be truly dangerous, while the harmful influence of the wind and dust is not prevented. These inconveniences can certainly be diminished to some extent by avoiding the crowded portions of the town and narrow streets, and by residing as much as possible on the southern side of the town. The climate, however, is in any case of but slight value.

Cairo should in reality be only recommended to patients of an extremely inactive temperament, whose constitutional condition is not unsatisfactory, the local lesions being slightly pronounced, and having given rise to no acute complication. Some inactive patients, the origin of whose disease is connected with scrofula or arthritis, perfectly answer to this set of indications. Even then, however, this station should not, in my opinion, be recommended unless the patients are able to interrupt the residence at Cairo by spending from four to six weeks upon the river Nile. This can be done in the most pleasant and comfortable manner, by means of boats known by the name of "Dahabeeyeh," which are let for this purpose, and during the whole period there is the advantage of a moist atmosphere, with more decided uniformity of temperature, so that the most harmful inconveniences of prolonged residence

* See note, p. 297.

in the town or its suburbs are thus avoided. This expedition, however, which can be of no real benefit unless performed in the above manner, is most costly, and since it is, in my opinion, indispensable, this to a great extent prevents recourse being had to the climate of Egypt. It should also be said that, even when the journey up the Nile is possible, this climate is one that requires the greatest attention and most serious consideration; for, like other places with decided qualities, it is never inert, having at one time a good, at another a bad effect.

The distinctive characters of the small group of patients to whom this climate is adapted have already been stated, but it should be remembered, in any case, that laryngeal complications of any gravity must be looked upon as an absolute counter-indication.

The reaction which has been termed indifferent occurs most frequently in practice, since the two preceding types are but rare, specially in their complete character. The cases which are included in the vague group of those with so-termed indifferent reaction may be looked upon as intermediate, in the sense of being equally distinct from those presenting the opposed forms of decided or undecided reaction, though perhaps in some of their features they may resemble one rather than the other form, numerous shades of difference existing which can only be mentioned. When the resemblance to one or other of the extreme types is so pronounced that neither its reality nor persistence can be doubted, it is necessary, in determining which form of climatic treatment should be adopted, to limit the selection to that group alone which is indicated by analogy, and to choose an individual station in the series which has been recognized as active or inert. It will be possible, however, since the reaction is not very pronounced, and if the other elements of decision permit such a course, to select from each group stations which have the least decisive action; and it is thus that Palermo,* Pisa,† and Pau,‡ should the reaction tend to become more pronounced, or Catania § in the reverse case, might be chosen.

In the truly intermediate group no indication is furnished by the individual reaction which is decidedly unpronounced, and the chief objects must be to prevent local complications and restore the constitution. The question of climatic treatment is thus unaffected by other conditions, the elements of decision having been already mentioned when the subject was generally considered; and according as the two fundamental indications are of equal or unequal importance should that climate be chosen which responds to the two indications, or to one of them in particular. The means of making this choice in different circumstances has already been considered, though one important remark should still be made.

An undecided reaction is most often of unstable character, the change towards torpidness or activity being always imminent. According to my experience, these two changes are not equally harmful, the alteration from undecided to pronounced reaction being in my opinion far more formidable than that of the reverse kind. Hence the patient should be removed from every influence which may promote this unfavourable change; and as regards the climatic effect in particular, it is more prudent and natural to attach these ill-determined cases to the group in which the reaction is decided, the climatic conditions being adapted to the individual conditions of the patient, and the stations which combine with their sedative action a more or less debilitating influence, namely, Pau and Pisa, being most unsuitable. Stations exist which are specially and, in my opinion, incontestably adapted to such cases, namely, Madeira and Algiers, and this character is due to their sedative or almost neutral character as regards

* See note, p. 328. † See note, p. 320.
‡ See note, p. 309. § See note, p. 328.

excitability, to their fortifying rather than debilitating effect, and to their thermal conditions, which are specially adapted to prevent the occurrence of local complications.

For analogous, though less clearly defined reasons, recourse may be had to the stations in Sicily, certain shades of difference existing which should be borne in mind.

Palermo has a less exciting effect than Catania on account of its different meteorological conditions, while the temperature, specially as regards variations during the day, and from day to day, is certainly more uniform. In addition to this, the vicinity of the two places, and the facility of communication between them, enables both to be utilized during the course of the same winter, experience having shown that this mode of treatment presents decided advantages. The individual excitability will determine the relative order in which the two stations should be employed; Palermo being first inhabited by patients of a somewhat excitable temperament, Catania by those in whom the reaction is unpronounced.

Lastly, two general remarks will be made of real practical importance. In the same way that laryngeal complications should be taken into serious consideration, intestinal symptoms necessitate special attention in the choice of climatic stations. If such symptoms are due to intestinal ulceration, they absolutely counter-indicate any change of residence; when, however, they are the expression of simple catarrh, or are due to some functional disorder of the stomach, these may be disregarded, unreservedly if the high climates are to be employed, but with certain precautions when the southern stations are in question. These precautions consist in the selection of those climates which are least apt to maintain or exaggerate the gastro-intestinal disorder. Even in the absence of previous pathological observations the question may be settled by the effects of acclimatization in the different regions. The places in which this usually occurs without disorder of the digestive system should be preferred in such cases; and according to my experience, notwithstanding that theoretically the opposite might be expected, Madeira in the first place, and then Mustafa and Sicily,* are at such times of special utility.

Secondly, the unfortunate consumptive patients should undertake no journey when they have reached the last stage of the complaint, and when the condition allows no hope of even temporary improvement. It is then too late for any climatic treatment to produce a beneficial result; the least that can result from a journey in these sad conditions is a temporary aggravation of the complaint, without any compensating advantage in the condition of the patient. But too often this aggravation is persistent, injudicious advice or mistaken tolerance only precipitating the fatal termination of the disease, and the patient dies in a mournful and isolated condition, far from his relatives and friends.

It will undoubtedly have been remarked that the different places of the Riviera † which adjoin the Mediterranean Sea have not been considered in the preceding discussion. The omission was not by mistake. Nor does it mean that they should not be utilized, or regarded as medical stations. As respects phthisis, which in this work is the sole disease under consideration, four places deserve such a title, namely, Cannes and Mentone in the French, San Remo and Spezia in the Italian Riviera. The reason for their omission is altogether different, and should be explained.

Notwithstanding the desiderata which they present with respect to public and private hygiene, notably as regards hydro- or aërotherapeutics, and which I only mention in the hope that this statement may produce some effect in causing them to disappear, the four places hold a most useful place in the large group of stations in the plain.

* See note, p. 321.   † See note, p. 324.

It may be remarked, indeed, that they offer, with respect to their climate, certain differences which correspond to the various kinds of individual reaction. Thus Cannes* and Mentone† have a more exciting effect than San Remo‡ and Spezia,§ the two former being therefore more appropriate to

* Cannes is a French winter station in the Riviera, belonging to the department Alpes Maritimes. See "Wintering in the Riviera," Wm. Miller; "Curative Effects of Baths and Waters," Julius Braun; "Health Resorts for Tropical Invalids," Moore; "Memoirs of Life and Work of C. J. B. Williams;" "Health Resorts," J. Burney Yeo, M.D.; "Principal Southern and Swiss Health Resorts," Wm. Marcet, M.D.; "Medical Guide to the Mineral Waters of France," etc., A. Vintras, M.D.; "Dictionary of Watering-Places," part ii. (L. Upcott Gill); "Handbook for Travellers to the Mediterranean" (J. Murray); "Handbook for Travellers in France" (J. Murray); "Dictionary of Mineral Waters," etc., B. Bradshaw; "Handbook for Travellers in Northern Italy," K. Baedeker; "Dictionnaire Encyclopédique des Sciences Médicales," Paris; "Guide to the South of France," C. B. Black; "The Climate of the South of France," C. J. Williams; "Influence of Climate on Pulmonary Consumption," C. J. Williams; "European Guide-book," etc., Appleton; etc.

† Mentone is a French winter station in the Riviera, department Alpes Maritimes. See "Curative Effects of Baths and Waters," Julius Braun; "Principal Southern and Swiss Health Resorts," Wm. Marcet, M.D.; "Wintering in the Riviera," Wm. Miller; "Dictionary of Watering-Places," part ii. (L. Upcott Gill); "Health Resorts," J. Burney Yeo, M.D.; "Handbook for Travellers in Northern Italy," K. Baedeker; "Handbook for Travellers to the Mediterranean" (J. Murray); "Handbook for Travellers in France" (J. Murray); "Wintering at Mentone in the Riviera," Alex. M. Brown; "San Remo and the Western Riviera," Arthur H. Hassall; "Winter and Spring on the Shores of the Mediterranean," J. H. Bennet, M.D.; "The Riviera," E. J. Sparks, M.A.; "The Climate of the South of France," C. J. Williams; "Influence of Climate on Pulmonary Phthisis," C. J. Williams; "European Guide-book," etc., Appleton; "Itinéraire Général de la France," Adolphe Joanne; etc.

‡ San Remo is in the Riviera, on the coast of the Mediterranean Sea, eastward of Mentone. See "Dictionary of Watering-Places," part ii. (L. Upcott Gill); "Visitor's Guide to San Remo," John Congreve; "Curative Effects of Baths and Waters" (Julius Braun); "Health Resorts for Tropical Invalids," Moore; "Health Resorts," J. Burney Yeo, M.D.; "Cities of Northern and Central Italy," Augustus J. C. Hare; "Principal Southern and Swiss Health Resorts," Wm. Marcet, M.D.; "Wintering in the Riviera," Wm. Miller; "Handbook for Travellers to the Mediterranean" (J. Murray); "Handbook for Travellers in Northern Italy" (J. Murray); "San Remo and the Western Riviera," A. H. Hassall; "The Riviera," E. J. Sparks, M.A.; "Climate of the South of France," C. J. Williams; "Influence of Climate on Pulmonary Consumption," C. J. Williams; "Guide to the South of France," C. B. Black; etc.

§ Spezia, or La Spezia, is a town in the Riviera, on the coast of the Mediter-

patients in whom the reaction is torpid or undecided, the latter to those in whom it is more pronounced. Nor is it only in pairs that the places differ in this respect. Each of them has its individual character, and it may be truly said that their exciting action diminishes gradually and with regularity from west to east. Thus it is that Cannes is more exciting than Mentone, Mentone than San Remo, San Remo than Spezia.

It is, in fact, clear that these stations may be of real benefit, and, as far as the comfort and improvement of the patients is concerned, no comparison can be made between the luminous winter of these regions, and the comparatively sad and gloomy condition of the towns in Central or Northern Europe. These places were not considered in the preceding discussion because, in my opinion, their effect is quite undecided, so that they cannot be of more than secondary importance. My belief is that these stations, like Pisa and Pau in the other series of places, should be looked upon as stations of reserve; that is to say, for the sake of greater precision and without any hostility to other places, as stations which may be employed when it is impossible to do better, the patients, it may be, refusing to make a longer and more difficult journey to climates whose action is more strongly qualified.

This condition of things should be well understood. The moment arrives when the condition of the patient no longer permits high altitudes and cold climates to be employed, when even the intermediate stations of the Tyrol and Switzerland, specially famed for their fortifying effect, are no longer appro-

---

ranean Sea, eastward of San Remo. See "Wintering in the Riviera," Wm. Miller; "Curative Effects of Baths and Waters," Julius Braun; "Principal Southern and Swiss Health Resorts," Wm. Marcet, M.D.; "Handbook for Travellers in Northern Italy," K. Baedeker; "Handbook for Travellers to the Mediterranean" (J. Murray); "Handbook for Travellers in Northern Italy" (J. Murray); "The Riviera," E. J. Sparks, M.A.; "European Guide-book," etc., Appleton; etc.

priate, and with respect to climatic treatment the southern stations are the only ones which can be employed. Does not this mean that the constitutional indication cannot be alone followed, and that it is as necessary, if not more so, to avert local complications? What, then, does the true interest of the patients require? It has been fully stated that the first requirement is the climate which most effectually prevents local complications, and without prejudice to the general indication. This condition is only realized in its entirety by climates whose properties are clearly expressed, specially as regards their meteorological uniformity and strengthening effect. Such is by no means the case in the Riviera,* since, however it is regarded, the same conclusion must be made as to its relative inferiority.

Even in making a comparison which must be completely artificial, on account of its unreality, and in considering these places in conjunction with the most southern stations of the group, and supposing for an instant that some distant analogy enables Cannes † and Catania,‡ Mentone § and Palermo,‖ San Remo ¶ and Algiers,** Spezia †† and Madeira ‡‡ to be classified together, the statement of this fact alone will justify the conclusion made. It shows, in fact, that though these places may be grouped together in pairs, no real similarity exists between them. Thus, if the climates be compared together as indicated above, or in any way which seems to be more rational, the four stations of the Riviera will invariably be found to have a lower mean temperature, with less uniformity during the day, from day to day, or from week to week. There is no hygrometric uniformity, since at Cannes, for example, the extreme numbers of 22 and 94, or even of 21 and 98, may be observed in the same month.

---

* See note, p. 324.   † See note, p. 393.   ‡ See note, p. 328.
§ See note, p. 393.   ‖ See note, p. 328.   ¶ See note, p. 393.
** See note, p. 297.   †† See note, p. 394.   ‡‡ See note, p. 243.

In the Riviera, again, there is more frequently wind, and the inconvenience of dust exists, which is by no means the case at Madeira or Mustafa, and scarcely so at Catania or Palermo. If, instead of joining these four southern stations together and comparing them with those in the Riviera, they are separately considered, as should certainly be done with the typical stations of Madeira and Algiers, the decision would be still more unfavourable, since their inferiority will be then found to be pronounced and absolute in every point.

Such being the incontestable results of this comparison between the climates, is it possible to regard them as at all of the same value, or to put such dissimilar places in the same group? In my opinion this cannot be done, and it is precisely the result of this comparative scrutiny which obliges me to separate the four places in the Riviera from the southern stations. Of the climatic conditions which are beneficial, if not necessary, some—as, for example, uniformity of temperature—are more fully realized elsewhere than in these places; while others—as hygrometric stability, the absence of wind and dust—are not at all realized therein. In consequence, though these may be good stations, they decidedly are not the best. In accordance, I believe, with other physicians, my opinion is that the best places must be sought; nor should we feel content with a good station unless no better one can be found. These should be reserved, as I think, for cases in which the patient refuses to utilize other places which have more characteristic effects, and which are at the same time more appropriate. It should be recognized that such cases are by no means rare; so that the stations in the Riviera, notwithstanding their relative inferiority, occupy, as they will probably continue to do for some length of time, a leading place in the climatic treatment of pulmonary phthisis.

As to the conclusion which finally resumes the preceding discussions, some hesitation is naturally felt in stating it, on account of the bad welcome which is always given to any truth which may disturb habits or threaten interests. This, however, must not prevail against the obligation of duty. The large number of medical stations, as was said at the commencement of this work, is more an appearance than a reality, and the conclusion, as will have been foreseen, confirms this statement in every point. Thus, during the whole period when high climates are beneficial, Davos, Samaden, and St. Moritz, at other times Madeira and Algiers in the first, and, though far removed as regards their value, Sicily, or exceptionally Egypt, in the second place, give the fundamental means of employing climatic treatment with the most certain and powerful effect. All other stations should be regarded as supplementary and accessory parts of treatment, which may be beneficial, but is certainly of less value from the fact that other places would be more so. This is a question of possibility, and half-measures, resulting from ignorance, routine, or prejudice, are always to be regretted, and only justifiable when it is impossible for patients to do more.

Such, then, is the conclusion of this discussion, which will not have been fruitless should it have shown that the disease is curable, that active therapeutic management is important, that the principles laid down are true, and the means to be employed with regard to the different forms of treatment advantageous. Salutary and practical notions will then be carried away from the perusal of this work, which will, in my opinion, repay the reader for the time and attention which have been devoted to the subject.

INDEX.

A

Absorption, fever of, 163, 164; treatment of pyrexia due to, 170, 171

Acromial extremity of clavicle, often not higher than the sternal extremity in phthisis, 83

Active, as contrasted with torpid, phthisis, 51

Acute miliary tuberculosis, 59, 60, 233–237

Adenopathy, cervical, bronchial, or mesenteric, the result of scrofulous phthisis, 11, 36

Aërotherapeutic treatment, 5, 102–4, 107, 108, 112, 113, 123, 124, 185, 191

Ætiology of phthisis, 64; opinion of R. J. Graves, 6; of Hughes Bennet, 6; of the author, 6–9

Age of parents, effect upon necessity of prophylactic treatment, 64

Aix-la-Chapelle, 259

Alcohol, produces sclerotic change, 160, 193; how to be administered, 190; beneficial when the fever of absorption exists, 193; employment of pneumonic phthisis, 226; a mode of administering with cinchona bark, 227

Alcohol, in pyrexia when to be reduced, 229

Alcoholism in parents, effect upon necessity of prophylactic treatment, 64

Algiers, 353, 369, 371, 374, 384, 391, 397

Allevard, 250; utility of mineral waters, 269

Alteration of voice, without appreciable cause, a presumptive sign of phthisis, 83

Altitude of place, guide in the selection of mineral water, 269; character of local lesions furnish no indication, 349; extent of local lesions important, 349; residence at high altitude counter-indicated by affections of the bronchial glands, generalized emphysema, abnormal conditions of the heart or large blood vessels, laryngeal ulceration, ulcerative endocarditis, nephritis, 350; its effect upon the pulmonary circulation removing the danger of congestion or hyperæmia, 293; effect on hæmoptysis, 293

Altitude, high, not to be suddenly inhabited, 253; counter-indicated by pronounced chlorosis or anæmia, 254

Amelie-les-Bains, 246, 275

Anæmia, persistent, a presumptive sign of phthisis, 83; if combined with scrofula, 114; preparations of iron beneficial in, 135; if pronounced counter-indicates the use of high climates, 254; of viscera, 291

Anapnograph (anapnographe), of Bergen and Kastus, 105

Animal food, when beneficial in the first period of phthisis, 126

Apex of lung, affinity of tubercle for, 88; inaction of as compared with rest of lung, 88

Arsenic, when beneficial, 116, 135, 144; how administered, 116, 145; its effects, 144; combined with cod-liver oil, 145; indications of non-tolerance, 147

INDEX.

Arthritic phthisis, 36, 280; clinical characters, 37; prognosis, 37; morbid anatomy, 37; case of, 38
Ascent in walking, beneficial in phthisis, 99
Atmosphere, its impression upon the respiratory tubes, 360
Atmospheric pressure, its effects, 290, 296, 297
Aufrecht, direction of clavicle in phthisis, 83, 84; pityriasis versicolor a sign of imminent phthisis, 86
Ausseé, residence at, 343, 346, 347, 351
Auto-infection, tubercular, its meaning, 65; mode of transmission of caseous products, 66

B

Bacteria, supposed connection with tuberculosis, 214
Baden, 280
Bagnères-de-Luchon, 278
Balsams, when to be employed, 156
Barèges, 260
Bath, 280
Baths, to be taken by those predisposed to phthisis, 93
Benzoate of soda, its inhalation, 6, 214, 215, 217, 218; disinfection of rooms by, 201
Bennet, Hughes, opinion respecting the ætiology and pathology of phthisis, 6
Bergen and Kastus, anapnograph (l'anapnographe) of, 105
Bert, P., opinion on respiration of compressed air, 104
Biel, researches upon koumiss, 132
Bismuth, employment in diarrhœa, 207
Bleeding, indicated in some forms of hæmoptysis, 205
Blisters, when applicable, 153
Blood-vessels, liable to rupture in pronounced anæmia, 257; disease of, 303
Boineau, examples of the transmission of phthisis, 74
Bollinger, researches regarding the infection of phthisis, 78
Bourboule, La, 246, 251, 261, 262, 277, 279, 281
Brandy, beneficial in some cases during the first period of phthisis, 127
Breathing, difficult, food in, 190
Bronchial adenopathy, in scrofulous phthisis, 36; counter-indicates residence at high altitude, 350
Bronchitis, modes of preventing, 89, 90; choice between, 90, 91; treatment of pyrexia in, 167
Broncho-pneumonia, resemblance to tuberculous pneumonia, 43, 44; treatment of pyrexia in, 167; cause of inflammatory fever, 189
Brünniche, mode of ascertaining inspiratory force, 85
Buhl, on tuberculous infection, 66, 68

C

Cairo, 328
Campfer, 318
Canary Islands, 297, 330
Cannes, 392, 395, 396
Carbolic acid, inhalation of, when advisable, 195, 196
Carbonic acid, in expired air, may indicate phthisis, 290
Caseation, its cause, 9, 55, 120; indication, 55
Caseous form of phthisis, 12; caseous degeneration of scrofulous foci a cause of phthisis, 65
Castan, examples of transmission of phthisis, 74
Catania, 328, 375, 386, 387, 391
Catarrh, peri-tubercular, 157
Catarrh of apex simulating phthisis, 124; preceding phthisis, 150
Catarrh of apex, at onset of disease an indication as regards high climates, 300; later in the complaint, 300
Catherine, St., 251
Caustics, employment of, 153
Canterets, 246, 250, 278
Cautery, employment of, 153
Cavern, appearances of, 18; parts surrounding, 19; cicatrix resulting from, 19; gravity of, 20, 22; indication of, 54; closure possible, 194; alcohol and creasote combined with inhalation of carbolic acid beneficial, 200; calcio phosphate beneficial, 200
Cervical adenopathy, 36
Challes, 263

INDEX. 401

Charcot, on identity of lesions in caseous pneumonia and the miliary granulation, 11
Chest, in phthisis, 82
Chlorosis, pronounced counter-indicates high altitude, 254
Chronic tuberculosis, 59
Churwalden, 318
Cinchona, when beneficial, 114, 227
Circulation of blood, effect of diminished atmospheric pressure on, 291
Classification of mineral waters, 245
Climates, their curative effect upon tubercle, upon phthisis, 275, 286; in each group, 354, 355, 368
Climates, high, precautions necessary in residing there, 255, 256; counter-indications, 349
Climates, division of in regard to treatment, 295; according to altitude, 295
Climatic conditions with altitude may confer immunity, 289; effect of temperature, 289; of winds, 289
Climatic treatment, importance of, 2, 263; guide in selection of mineral water, 269
Clothes of children liable to phthisis, 93
Cod-liver oil, when to be replaced by glycerine, 191; when and how to be taken, 136-140
Complications, prevention of, 118; treatment, 118
Compressed air, respiration of, 103, 104
Condition, general, of patient, 361
Confinement to the house, effect in phthisis, 90; to be deferred as long as possible, 201
Congestion, intercurrent, effect of, 54
Conjunctival hæmorrhage, liable to occur in pronounced anæmia, 257
Consanguineous marriages, effect of, 64
Constitutional condition, importance of, 54
Constitutional debility, indication of treatment, 121
Corfu, 375, 387
Corning, experiments of regarding the infection of phthisis, 76
Cough, and its treatment, 156; attending inhalation of carbolic acid, 197
Counter-irritation, when advisable, and its effects, 153, 190, 204, 208
Country air, advantage of in phthisis, 95

Cowshed, milk to be taken in during the first period of phthisis, 127; atmosphere of beneficial, 127
Creasote, why and how to be employed, 156, 191, 194, 195, 207
Cube, example of cure of pneumonia of the apex, 124
Cupping-glasses, when and how to be employed, 190
Curability of phthisis, 11, 22-24, 27, 29

D

Davos, 287, 317, 324, 332, 334, 335, 338, 341-345, 348, 351, 397
Deep position of clavicle in phthisis, 84
Diabetes mellitus, cause of phthisis, 32; in parents imposes necessity of prophylatic treatment, 64
Diabetic phthisis, 38
Diarrhœa, connected with dyspepsia, treatment, 207
Digitalis, not to be administered in hæmoptysis, 212; in pneumonic phthisis, 227; indication of, 231, 232
Diminution of atmospheric pressure, effects of, 290
Diphtheria possibly followed by broncho-pneumonia, 44; by phthisis, 71
Disinfection, modes of practising in phthisis, 81, 201
Dittrich, on tuberculous infection, 66, 68
Dress, most suitable in phthisis, 95
Duality, clinical, of the disease, 11, 12
Duckworth, on tuberculous infection, 68
Dührssen, on respiration of compressed air, 104
Duke of R., report of case, 22
Dwelling, most suitable in phthisis, 95
Dyspepsia, persistent, a presumptive sign of phthisis, 83; not usual at the onset, 126; treatment, 126; attending the first formation of tubercles, 185; treatment, 185
Dyspeptic diet in phthisis, 205

E

Eaux bonnes, 260, 278
Egypt, 297, 321, 353, 364, 365, 387-389

Emaciation, 50
Engadine, the, rarefied atmosphere, 256; absence of phthisis, 286; effect of, 317; residence in, 332, 341, 342
Enteric fever, possible cause of phthisis, 71
Emphysema, surrounding cicatrix of previous cavern, 19; pulmonary, counter-indicates the employment of high altitude, 253, 303, 352
Ems, mineral waters of, 272
Enteritis, ulcerative, counter-indicates residence at high altitude, 350
Epigenesis, doctrine of, 8
Epistaxis, liability to, 257
Erethism, 46, 51
Ergot of rye, its employment in hæmoptysis, 211
Ergotine, beneficial in hæmoptysis, 209, 212
Exciting climate, effects of, 365; distinguished from fortifying, 366
Exercise, muscular, necessary in those predisposed to phthisis, 94; beneficial in phthisis, 99
External appearance of those predisposed to phthisis, 82
Extremities of fingers in phthisis, 82
Eyes, in phthisis, 82

F

Falkenstein, 331, 343, 346, 347, 351
Fatty substances, to be avoided in phthisis, why, 126
Fever, ideas of author respecting, 9; intermittent, indicating existence of tubercles, 189; inflammatory, 189, *See* Pyrexia.
Fibrous tissue, development of, 55
Fibrous transformation of tubercle, 19–21, 30, 32
Figure in phthisis, 82
Fischl, on diminution of inspiratory force in phthisis, 84
Fleming, on the infection of phthisis, 78
Flindt, on the infection of phthisis, 76, 77
Fluela pass, temperature at summit of, 337
Food, necessary to those predisposed to phthisis, 94; most suitable in phthisis, 96; in pneumonic phthisis, 225
Foot, on the infection of phthisis, 78

Forges-les-eaux, mineral waters at, when beneficial, 251, 252
Fortifying treatment, its value, 91; as distinguished from tonic, 363, 364; from exciting, 366
France, south of, watering-places in, 297

G

Gallic acid, beneficial in hæmoptysis, 208
Gastric catarrh, treatment of in phthisis, 203
Gastro-intestinal functions, effect of disorder in, 46; guide in selection of mineral water, 269
Gaudal, residence at, 331
Geneva, places at eastern extremity of, when beneficial, 320
Gerlach, researches of regarding the infection of phthisis, 78
Giddiness, produced by inhalation, 197
Glandular enlargement, treatment of, 117; importance and danger of, 262; La Bourboule beneficial, 262
Glycerine, employment of in the place of cod-liver oil, 139–143, 191
Gorbersdorf, protective altitude of, 288; residence at, 331, 343, 346, 347, 351
Grancher, work on identity of lesion in caseous pneumonia and the miliary granulation, 11
Graves, R. J., pathology of tubercle, 6
Groups of climates, when applicable, 368; types of, 369
Gurnigel, mineral waters, when beneficial in phthisis, 250, 260, 279

H

Hæmoptysis, 47, 48, 50, 71, 72, 110, 111, 134, 208, 212, 213, 272, 308, 349
Hæmorrhagic purpura, 257
Hair, in phthisis, 82
Hünisch, on direction of clavicle in phthisis, 84; registration of inspiratory movements, 85
Heart disease, counter-indicates the employment of high altitude, 253, 303, 350
Heidenhain, experiments of, 341
Hereditary form of phthisis, 30, 31, 63
Herpetic phthisis, 38, 39, 280
Heteromorphism, doctrine of, 8

INDEX. 403

High climates, precautions requisite before patients reside there, 256, 310, 350; utility and counter-indications of, 253, 256, 287, 300, 301, 302; effects on the circulation, 293, 294; treatment by, 351
Hughes Bennett, pathology of tubercle, 6
Huguenin, on tuberculous infection, 66
Hutchinson, J., spirometer of, 105
Hydropathy, treatment by, 5, 122, 191
Hygienic treatment, 5; of secondary phthisis, 32; during premonitory period, 117; during apyretic phases of confirmed disease, 121; hygienic condition of region should be known, 327
Hygrometrical condition of place, importance of, 359

I

Ice, application to chest wall in hæmoptysis, 213
Iceland, immunity from phthisis in some parts, 286
Inaction of apex, as compared with other parts of lung, 88, 107
Indifferent reaction, 389–391
Indoor life, diminishes good effect of altitude, 288
Infection, possible transmission of phthisis by, 73, 74, 80
Inflammation, intercurrent, effect of, 54
Inflammatory affections of bronchial tubes, supposed result of cold air, 340; not if dry, 341
Inflammatory fever, 189
Inhalation of iodine, turpentine, tar, carbolic acid, 195–199
Injections, sloughing of skin caused by, 182; antipyretic effect, 183
Innate phthisis, meaning and form, 32
Inoculation, transmission of phthisis by, 80
Inspiratory force, diminished in phthisis, 84; mode of ascertaining, 85
Intercurrent congestion or inflammation, effect of, 54
Intermittent fever, indication of tubercles, 189
Intestinal ulceration, counter-indicates high or unseasonable climates, 306

Intestinal complications, not to be left unconsidered, 382, 391; treatment of, 392
Iodine, inhalation of, 195; tincture of, its employment, 207
Iron, when beneficial, 115, 134; parchloride of, injection in hæmoptysis, 212; inhalation, 212
Iron points, heated when beneficial, 183
Isch, 272
Italy, some watering places of, 297

J

Jacobson, opinion on respiration of compressed air, 104

K

Klebs, researches of with respect to infection, 74
Knævenagel, cases of auto-infection from pleurisy, 69
Kommercil, researches of with respect to infection, 78
Koumiss, its origin, employment, etc., 129–132
Kreuznach, mineral waters of, 261, 262
Kroczak, treatment recommended by, 214, 215
Kuchenmeister, first recommended use of mountain climates in winter, 333

L

Laennec, ideas of as to the origin of tubercle, 8; incurability proclaimed by, 28
Laryngeal complications, not to be neglected, 382, 391
Laryngeal irritation, causes, treatment, etc., 155
Laryngeal ulceration, counter-indicates high and rigorous climates, 306, 350
Laryngites, chronic, its effect, 47
Laryngo-bronchitis, a complication not to be neglected, 117
Laudanum, employment in diarrhœa, 207
Lazarus, on respiration of compressed air, 104, 106; example of cure of pneumonia of apex, 124

Lebert, on tuberculous infection, 66, 68
Lesion, pulmonary, effect of as regards residence in a high climate, 306, 307
Leube, on transmission by means of milk, 78
Liebig, effect of breathing compressed air, 105
Lippspringe, 281
Local disease, preservation from, 356
Local treatment, how and when to be employed, 152
Lochmann, researches of regarding infection of phthisis, 78
Louëche, 281
Lucas-Champonnière, spray producer of, 195
Luchon, 259
Lugano, 347, 380, 381

M

Madeira, 297, 353, 365, 369, 371, 373, 384 891, 392, 397
Malnutrition, meaning of, 54; cause of tuberculosis, 55, 61, 88, 39; indication of treatment, 121, 156
Marriage of those predisposed to phthisis, 101
Mazzotti, cases of tuberculous infection, 68
Meals, to be taken during the first period of phthisis, 127
Measles, possibly followed by broncho-pneumonia, 44; possible cause of phthisis, 71
Meat, possible transmission of phthisis by, 81
Menstrual disorders, if persistent, a presumptive sign of phthisis, 83
Mental studies, not prevented by treatment, 100
Mentone, 395, 896
Meran, 325, 347, 353, 366, 369, 380, 381
Mesenteric glands, affection of, 36
Meteorological conditions, no guide in selection of mineral water, 269
Meteorological numbers, result of personal observation, 370
Miliary tuberculosis, 39, 42, 189; acute, 60
Milk, of cows affected by tuberculosis, a cause of infection, 78, 80; beneficial in first period of phthisis, 127;

how and where to be taken, 127; may be mixed with alcohol, 129
Mineral waters, their use, 5, 117, 229–243; classification, 245; choice of, 246, etc.; counter-indications to, 263–266; selection of, 269; no effect on tubercles in primary phthisis, 270; division of, 271; utility of, 281, 282
Mogador, 378
Mont Doré, 277, 280
Montreux, 325, 347, 366, 369, 381
Morocco, 297, 330
Morton, doctrine of, with regard to septic air, 72
Mosso, effect of breathing compressed air, 106
Muscular exercise, necessary in infancy to those who are predisposed to phthisis, 94
Mustafa Supérieur, 365, 392

N

Nauheim, 261
Neck, in phthisis, 82
Nephritis, counter-indicates residence at high altitude, 350
Nervo-vascular excitability, 303
Neukomm, opinion on respiration of compressed air, 104
Neurotic conditions, associated with constitutional debility, 344
Norway, 287
Nutrition, insufficient, origin of the tubercular diathesis, 9

O

Opium, beneficial in hæmoptysis, 208
Ordinary form of phthisis, chs. vi., vii., viii.
Origin of phthisis, varieties dependent on, 30
Orth, on tuberculous infection, 66, 68
Ou——, Prince, case of, 48
Oxygen, quantity in air breathed, possible effects of, 290

P

Palermo, 328, 371, 375, 376, 384, 390, 391
Panticosa, 276

Pau, 309, 320, 366, 371, 385, 390
Peptone, use of in advanced phthisis, 205
Personal knowledge of places requisite, 2
Perchloride of iron, beneficial in hæmoptysis, 209; injection, 212; inhalation, 212
Phthisigenic pneumonia, 13
Phthisis, ordinary form of, 379, ch. vi., vii., viii.
Pigmentation, from diminished atmospheric pressure, 291
Pisa, 320, 366, 371, 376, 384, 390
Pityriasis versicolor, development of said to be a sign of imminent tuberculosis, 86
Plain, stations of, 319, etc., 323
Pleurisy, a cause of auto-infection, 69; cases recorded by Knævenagel, 69; persistent, 344
Pneumatic treatment, 5
Pneumonia, as a complication, not to be neglected, 118; of apex simulating phthisis, 124; treatment of pyrexia caused by, 167; counterindicates high climates while pyrexia exists, 307, 308; chronic, 344
Pneumonic form of phthisis, 12, 26, 371; prognosis, 42, 43; onset, etc., 219-226, 232, 233; mineral waters in, 281
Ponfick, on tuberculous infection, 66
Portugal, 297
Position of clavicle, 84
Primarily acquired phthisis, 34
Prognosis, upon what dependent, 55, 120
Prophylactic treatment, chs. iv., v.
Protection from northern winds at Davos, Samaden, and St. Moritz, 335
Purity of air at Davos, Samaden, and St. Moritz, 335
Purpura, hæmorrhagic, 257
Pyrexia, attending hæmoptysis, 48, 49; significance of, 51, 120, 161-3, 167, 184, 194, 204; treatment of, 186, 187, 229-232; counter-indicates change of residence and high climates, 305
Pyrmont, 252

Q

Quinine, its employment, 165-167, 169, 190, 211, 212, 230

R

Ragatz, 255
Rarefaction of the air, effect on respiration, 292; on nutrition, 292, 293; on caverns, 307; at Davos, Samaden, and St. Moritz, 334, 335
Reaction, in hydropathy, its persistence necessary, 98; its triple quality, 123; character of, 266, 383; indifferent, 389-391.
Receptivity, law of, 80
Recovery from phthisis possible, 20, 21; case of, 22
Reich, on infection of phthisis, 76, 77
Residence, country, selection of, 100; not to be in a town, 327; of patients affected by phthisis residing together, 339, 340
Rhatany, beneficial in hæmoptysis, 208
Riding, beneficial in phthisis, 99
Riviera, residence in the, 353, 365
Rokitansky, plan of treatment recommended by, 214, 215
Routine, influence of, in selection of climates, 283
Royat, 272
Russia, lowly situated regions of northern, frequence of phthisis in, 287

S

St. Catherine, mineral waters of, 251, 252, 257
St. Gervais, 281
St. Honoré, 275
St. Moritz, mineral waters of, 250, 252, 332; climate of, 252, 335, 336, 342, 397
Salicin combined with creasote, their employment, 195
Salicylate of soda, when to be employed, 174, 193; its effects, etc., 178-181
Salicylic acid, as a febrifuge, 5; how and when to be used, 165, 166, 174, 192-195, 207, 226, 230; its effects, 176, 193
Salins, 251, 262
Samaden, 324, 332, 335, 342, 351, 397
San Remo, 395, 396
Sanitary conditions of place, 269, 326
Saxon, 261, 279
Schirmunsky, researches of, 106
Schinznach, 260, 281

Schnapf, spirometer of, 105
Schnitzler, on respiration of compressed air, 105
Schottelius, experiments regarding the infection of phthisis, 76
Schreiber, experiments regarding the infection of phthisis, 76
Schwalbach, 251, 252, 258
Sclerosis following catarrh, 157
Scrofula, a cause of phthisis, 33, 35; effect upon treatment, 64; caseous degeneration due to, 65; combined with anæmia, 114; and ordinary phthisis, 135; combined with constitutional debility, 258; treatment of, 258
Scrofulous phthisis, 33, 35, 36
Sea voyages, 101
Secondary phthisis, 35
Semma, experiments regarding the infection of phthisis, 74, 79
Shower-bath beneficial in phthisis, 96, 97
Sicily, 321, 353, 391, 392, 397
Sick and Anderson, facts on acute miliary tuberculosis reported by, 234
Silesia, 287
Skating at Davos, Samaden, and St. Moritz, 342
Skin in phthisis, 82
Sloughing of the skin caused by injections, 182
Soda, chlorinated, 272; chlorinated bicarbonate of, 272
Soden, 272
Sodic chloride, 258
Sommerbrodt, case of pneumonia of apex, 124
Soyka, cases of tuberculous infection, 67
Spa, mineral waters of, 251, 258
Spain, 297
Spengler, doctrines of, 333
Spezia, 395, 396
Stations, how to procure knowledge of, 361, 362
Stembo, researches of, 106
Stomatitis, 228
Styria, 281, 288
Suckling of child, not to be carried on by mother affected with phthisis, 92; case of, 102
Sulphurous waters, beneficial in scrofula, 258; their utility, 275
Sun, length of exposure to, at Davos, Samaden, and St. Moritz, 335

Sweden, 287
Switzerland, 287, 296
Symptoms at onset of pulmonary phthisis, 120

T

Tangiers, 378
Tappeiner, experiments of on the infection of phthisis, 75
Tar, its employment, 156, 195
Tartarated antimony, its employment, 188
Teeth in phthisis, 82
Temperature, uniformity of, important, 357
Teneriffe, 377
Thapsia, plasters of, when advisable, 154
Thaon, work on identity of lesions in caseous pneumonia and miliary phthisis, 11
Thompson, opinion of as to hæmoptysis being a cause of phthisis, 72
Tobogganing, at Davos, 342
Torpid phthisis, 51
Torpid reaction, 337
Transformations of tubercle, 14, 15, 18; fibrous ditto, 19-21, 30, 32
Transmission of phthisis, 63, 80
Treatment, effect upon curability of disease, 57, 58; prophylactic, chs. iv. v.; of phthisis at onset, 120; of ordinary phthisis, 147; of pneumonic phthisis, 232, 233
Tubercle, production of, 149
Tuberculosis, chronic, 59, 348; pneumonic, 59; acute miliary, 59, 233-237
Tuberculous pneumonia, 44; hæmoptysis in, 50; deposits, effects of, 192
Turpentine, its employment, 156, 195
Types of four series of climates, 369
Typhoid fever, possibly followed by broncho-pneumonia, 44
Tyrol, 296

U

Ulceration, pyrexia due to, 167, 168
Uniformity of temperature, preserves from local disease, 357
Uriage, 251

V

Ventilation of rooms at Davos, Samaden, and St. Moritz, 338, 339
Verga, experiments of with regard to the infection of phthisis, 74
Vienna paste, application of, 154
Villemin, experiments of with regard to the infection of phthisis, 74
Virchow, on tuberculous infection, 66
Vital capacity, 105
Voice, alteration of when a presumptive sign of phthisis, 83
Vomiting, treatment of, 206; produced by cough, 206

W

Waldenburg, observations on the effects of breathing compressed air, 106
Walking exercise, when beneficial in phthisis, 191
Warm climates, 283
Wasting period of disease, excludes high climates, 305
Weissenburg, 273, 274
Wildegg, 261
Winds, 269, 358
Wine, beneficial in the first period of phthisis, 126, 127

www.ingramcontent.com/pod-product-compliance
Lightning Source LLC
Chambersburg PA
CBHW030557300426
44111CB00009B/1010